BOOMERANGS

*How to Make and
Throw Them*

BOOMERANGS

How to Make and Throw Them

BERNARD S. MASON

DOVER PUBLICATIONS, INC.

NEW YORK

Published in Canada by General Publishing Company, Ltd., 30 Lesmill Road, Don Mills, Toronto, Ontario.

Published in the United Kingdom by Constable and Company, Ltd., 10 Orange Street, London WC 2.

This Dover edition, first published in 1974, is an unabridged and unaltered republication of Part I, "Boomerangs," from *Primitive and Pioneer Sports* by Bernard S. Mason, as originally published by A. S. Barnes & Company in 1937.

International Standard Book Number: 0-486-23028-7
Library of Congress Catalog Card Number: 73-94346

Manufactured in the United States of America
Dover Publications, Inc.
180 Varick Street
New York, N.Y. 10014

TABLE OF CONTENTS

CHAPTER		PAGE
I.	Boomerangs the World Around	3
II.	Making Cross-stick Boomerangs	11
III.	Pin-Wheel Boomerangs—How to Make Them	36
IV.	Boomabirds—How to Make Them	48
V.	Tumblesticks—How to Make Them	58
VI.	Australian Boomerangs—How to Make Them	64
VII.	Cardboard Boomerangs	72
VIII.	How to Throw Boomerangs	77
	Index	99

BOOMERANGS

*How to Make and
Throw Them*

CHAPTER I

Boomerangs the World Around

There is mystery in the boomerang. Nothing more than a piece of wood, yet with uncanny accuracy it circles through the air and comes back to the thrower. Obviously enough the thrower holds no magnet that plays upon the sailing missile, has no magical·power that pulls it back to him—the secret rests entirely in the mechanical construction of the stick and the physical laws of ballistics. Yet, however completely these laws may be expounded and comprehended, there will always be something of magic in the uncanny ability of these flying sticks to find their way into the waiting hands of the thrower.

Doubtless this element of mystery has much to do with the universal appeal of the sailing boomerang. Certain it is that it holds a peculiar fascination to young and old of every clime. There is intrigue in the very appearance of the boomerang in the air, in the ease and grace with which it soars and sails, circles and double circles, and finally floats effortlessly back to its starting point. No bird ever soared the sky with more harmony of movement and easy elegance than does a good boomerang.

And then, too, coming as it does from the primitive, boomerang throwing carries with it that glamour that all things primitive have to the civilized mind.

Little wonder that the thrower takes so much pride in sailing these graceful sticks. The sheer beauty of the flight and the perfection and precision of the return are a never-ending joy—he throws until weary and obtains thereby excellent physical exercise and the self-expression that comes from good recreation. Almost equally as great is the appeal to the spectator: one glimpse of the floating missile and the onlooker is arrested—he remains to thrill and to enjoy æsthetic satisfaction.

Boomerang throwing has few equals as an individual sport. It offers as much of physical benefit as any throwing and catching activity. It grips and compels. It is one of those colorful activities that not only challenges the performer but has unique show value. It carries a romantic appeal—it does things to the imaginations of people. Being an unusual type of activity, it serves admirably as a hobby. I have never known a person to become familiar with the art of the boomerang and not be caught in the irresistible, sweeping tide of its appeal. The sport becomes a major interest and a constant source of pride.

But there is another aspect of this sport that plays a most conspicuous rôle in its appeal—*the making of the boomerang is as interesting as the throwing*. In fact the making and the throwing are inseparably related in the full enjoyment of the pastime. There is pleasure in throwing a boomerang that is purchased or obtained from some one else, but it is in no respect comparable to the joy and thrill that results in handling one which you yourself have made. All the time the boomerang is being whittled you are looking forward to throwing it—constantly in your mind is the question, "Will it come back?" And when the last chip has been removed, you hasten to hurl it—*and it works!* There is thrill and glowing satisfaction as can come from few other pastimes! Even the old-timer at the boomerang game never fails to experience it; he may have made a thousand boomerangs, yet each time he throws a new one and it works perfectly just as he planned that it should, he feels a surge of pride and satisfaction that is worth many times over the effort required for the making. It is a feeling of craftsmanship, of having been the cause! So great, in fact, is this joy that comes from seeing a newly-made boomerang work perfectly, that one is always tempted to put the stick away after throwing it enough to test it thoroughly, and then to make another designed to act in a different way.

And happily, the making is easy—nothing more forbidding than jackknife whittling, and not very difficult whittling at that. Most boomerangs are made from soft wood that is very easy to work. And when a boomerang is roughly whittled out, so anxious is the maker to see if it will function that he throws it before finishing it carefully. If it works, he is usually reluctant to touch it again with the knife and so it is left in a crude and rough state. But this crudeness enhances rather than detracts from its appearance. Since these are primitive instruments, a machine-like perfection in appearance unmakes the picture. All of this adds to the simplicity of the making.

PLACE IN THE EDUCATIONAL AND RECREATIONAL PROGRAM

The recreational value of boomerang making and throwing is clearly indicated in the foregoing paragraphs.

An ideal play activity serves two purposes: It is gripping and compelling at the moment—it appeals as play, and makes the typical contributions of play to growth—and secondly, it is an activity of the type that will carry on throughout life, that is, the learning of the skills necessary to play it is an education for leisure.

The use of boomerangs meets these requirements outstandingly. It is an individual rather than a team or group activity; experience indicates that the activities that can be enjoyed alone or with two or three others are the life-time type. Once one is familiar with the art of the boomerang, he will doubtless come back to it again and again throughout life—he is equipped with an enduring and lasting hobby. This being the case, the use of boomerangs has a place in any school or club that attempts to educate in leisure-time activities.

In any group or class of children, there will be some who for temperamental or physical reasons do not participate regularly in the usual team games. Boomerangs will hold a peculiar appeal to such types. In individual corrective work in physical education, this activity has much to recommend it. But its glamour is not confined to these special types of people—practically all boys, girls, and men will be challenged by the boomerang in any play and recreational program.

Viewed from the handicraft standpoint, boomerang making has a purpose, an objective unknown in the usual crafts—it has a reward beyond the satisfaction of good craftsmanship—that of

throwing the boomerang when completed. This activity is therefore ideal for use in clubs, playgrounds, camps, and schools, both
as a craft and as a sport. In schools the making of boomerangs
is usable in manual-training and industrial-arts departments, and
the throwing of boomerangs is delightful in the physical education department. Most boomerangs of the types we shall recommend work more effectually indoors where there are no air
currents to interfere and consequently, the school gymnasium is
an ideal place for their use.

The element of danger immediately comes to mind when
boomerangs are mentioned, and this fear has often militated
against the consideration of the activity in school manual-training departments. The element of danger has been greatly exag-
gerated, due in part to the fact that few are familiar with types
of boomerangs other than the curved Australian type. This
boomerang is heavy and does carry a considerable element of
danger. Not so, however, with the come-back sticks which will
be described in the chapters that follow. The use of the curved
Australian boomerangs is not recommended for boys. There are
other types which are light and much more efficient, operate in
a small space, and are relatively safe. Given instruction equal
to that afforded for other sports, the element of danger can be re-
duced to the point where it is no more important than in any of
the usual activities of children.

TYPES OF BOOMERANGS

The art of the boomerang, both in respect to the methods of
construction and the manner of throwing, is practically unknown
as far as the general public is concerned. The secrets of the
boomerang have been held in this country by a very few actors
and circus performers, and even these can be counted on the
fingers of one's hand. Many of the boomerangs described in
the chapters which follow are entirely original in design.

To the average person, the word *boomerang* brings to mind
the curved weapons used by the Australian primitives. In fact,
Webster's dictionary defines the word as "a curved or angular

club used, mainly by the natives of Australia, as a missile weapon. It can be thrown so that its flight will bring it back near to the place where it was thrown."

For the purpose of this book it is assumed that the name, *boomerang*, is applicable to any missile which when hurled will return to or near the place from which it was thrown. With this as a definition, there are many types of boomerangs in addition to the curved style used by the native Australian. In fact the curved Australian type (Figure 34, page 66) is the least efficient type as a come-back stick. There are of course excellent and perfect boomerangs of the curved type, but they are few in number and as a type they are not comparable in efficiency to the others that these chapters will describe. No boomerang is worth the name if one has to step to reach it as it returns.

In addition to the curved or Australian style, there are three main types, each with many variations: (1) *The Cross-stick Boomerang*, (2) *the Boomabird*, and (3) *the Tumblestick*.

The *Cross-stick Boomerang* consists of two or more sticks fastened together—Figure 4, page 19, shows the appearance of a boomerang of this type. The methods of construction and the many varieties are described in Chapter II.

The *Pin-wheel Boomerang*, which, in fact, is a variation of the Cross-stick, is made of three sticks as shown in Figure 18, page 37. It is the ·most popular and efficient of the boomerangs. Chapter III describes it in detail.

The *Boomabird* is a boomerang so constructed as to look like a bird. This novelty, shown in Figure 26, page 51, is unique both in appearance and performance—it does many fascinating and clever things in the air. The Boomabird and its variations are described in Chapter IV.

The *Tumblestick* is in many respects the most unusual type of boomerang. It is essentially a straight stick that will return to the thrower. Figure 31, page 59, shows it and the methods of making and throwing are presented in Chapter V.

There are many other styles of boomerangs described in the

following chapters but each is related to one of the four main types mentioned above.

Although the art of the boomerang was common in many parts of Australia, it was a single tribe of Australian primitives living in New South Wales that applied the name *boomerang* to the come-back missiles they used in hunting, the other tribes using other names for these same weapons. The word *boomerang*, however, has become universally accepted in the English language. The Australian word *womera* is occasionally seen in print as synonymous with *boomerang*, but this is incorrect— womera refers to spear throwing.

It should not be assumed that boomerangs were the exclusive property of the primitive Australians. The ancient Egyptians are said to have made extensive use of a boomerang-like missile, and today there are certain sections of northwest Africa in which returning weapons resembling the Australian type are still used. Furthermore, the natives of South India use a boomerang-shaped weapon made of ivory and steel which can be made to return in the direction of the thrower. The Hopi (Mosquis) Indians of Arizona use a type of boomerang resembling the Australian for hunting, this being the only record of the use of boomerangs on the North American continent.

There are two types of boomerangs used by the Australian Bushmen—*the return boomerang* and the *non-return boomerang*. The names used in this classification are scarcely correct, however, in that both can be thrown so as to return toward the thrower, although they must be thrown in different ways. The return type, when held in a vertical position and thrown straight forward will circle to the left and return. If the non-return type were thrown in this way, it would go straight forward with great speed and accuracy in the direction of the target at which it was hurled, showing not the slightest inclination to turn either to the right or left. However, if one of these non-return boomerangs is held parallel to the ground and thrown, it will rise high in the air and then volplane down to the ground near the sender.

Much of myth has been said and written regarding the use of the boomerang by the Bushmen. For example, we hear the boomerang glorified as a weapon of warfare. Certain it is that the non-return boomerang would have been a valuable and efficient fighting weapon, for it is probably true that the Bushmen had no other weapon that could be hurled so accurately for so great a distance. However, these boomerangs were so difficult to make, the materials so hard to find, and there were so few really expert boomerang makers, that it is doubtful that they saw much service in battle. Spears and clubs did good enough service in the close fighting that primitives enjoy and were much easier to fashion. A fine boomerang was so highly prized for hunting that its owner would be reluctant to risk losing it in warfare—and lose it he probably would, sooner or later, for when hurled so as to go straight for the target, the boomerang would not return to the sender, either if it missed its mark or if it hit it. But in spite of all this, there is no doubt that boomerangs did fighting duty, although not to the extent that is popularly supposed.

Boomerangs were chiefly hunting weapons. Here again, there were limitations to their use contrary to the common conception. It is true that the non-return type was used in hunting the kangaroo, and there is no gainsaying that a three or four-foot boomerang would easily cripple an animal of that size. True it is, too, that it would zip toward its prey with amazing speed and accuracy. However, such use in hunting animals was incidental to its use in hunting *birds*. The method here was not to throw at a perched or standing bird so much as it was to hurl the boomerang into a flock of flying birds. When the non-return boomerang is thrown horizontally it rises high into the air, and if sent into a flock of flying birds, its arms, whirling about viciously as they do, would have an excellent chance of bringing meat into the lodge. Curiously enough, birds are attracted to flying boomerangs, rather than being frightened away from them. The appearance of the boomerang in the air resembles a flying bird so much as to serve as a sort of decoy.

While the smaller and lighter boomerangs of the return type were also frequently used in hunting birds, they found their greatest use as playthings in sport. If such a boomerang were thrown at an animal or bird and missed its mark, it would circle back toward the sender. However, if it hit the object, it would not return, but rather drop to the ground.

While boomerangs have been used elsewhere in the world, it was certainly on the Australian continent that they found their greatest development. The flat country that is the home of the Bushmen is ideal for the manipulation of these curved sticks and the little brown men of that region who were so adept in their use relied heavily on them for meat. The Bushmen made upwards of twenty different styles of curved boomerangs, ranging from ten inches from tip to tip, up to four-and-one-half feet in length.

The large non-return type of boomerang is of little interest in recreation. In appeal it gives way rather completely to the smaller models of the return type. A boomerang that circles around to the left and comes back to you is always more interesting than one that merely sails up and then skids down again.

The difference in construction of these two types of boomerangs is described in Chapter V.

CHAPTER II

Making Cross-Stick Boomerangs

The Cross-stick Boomerangs are the pinnacle of boomerang perfection. They are the most accurate of all the boomerangs and consequently are the most satisfying. So accurate is a good Cross-stick Boomerang that an expert can stand upon a stage, even in a small theater or hall, and throw it out over the heads of the audience without fear that it will hit a wall or dive surreptitiously onto the head of an unsuspecting spectator. So accurate is it in fact that one can often "call his shots," that is, throw the boomerang and immediately hold out his hand indicating the exact spot to which it will return. The symmetrical, balanced construction of these boomerangs causes them to cut a more perfect circle in the air and return with more precision than any other type. A good Cross-stick or Pin-wheel can be depended upon to act in precisely the same way each time it is thrown.

The Cross-stick Boomerangs are usually made of light, soft wood, and this, together with their accuracy and dependability, recommends them as relatively harmless, delightful playthings. They can be used with safety in any gymnasium or large hall, following the directions given in Chapter VII, "How to Throw Boomerangs."

Moreover, the Cross-sticks are the easiest type of boomerang to make. The process is so simple that any one, after reading the instructions in this chapter, should be able to fashion one in a few minutes that will work with the utmost perfection if properly thrown. Certainly the beginner at the sport should be introduced to the simple Cross-stick before he makes the acquaintance of the other styles. In fact, practically all of the essentials in the art of the come-back missile are presented in this chapter on the Cross-sticks—this is the basic type. The unique models of boomerangs described later are unusually fascinating novelties and very much worth while, but they are merely colorful variations based on the principles of the Cross-stick.

There are two main types of Cross-sticks: The first is the *Four-wing Cross-stick* made of two pieces of wood set at right angles to each other as in Figure 4, page 19. This is the simplest form of boomerang. The second is the *Pin-wheel* which is a six-wing boomerang made of three pieces of wood fastened together in the middle. It is illustrated in Figure 18, page 37. This is the Cross-stick at its best—the steadiest and most dependable of the boomerangs. The present chapter deals with the simple Four-wing Cross-stick and its variation, the two-wing type. The Pin-wheels are described in the next chapter.

The methods of throwing boomerangs are described in Chapter VIII. Since all boomerangs are manipulated according to the same principles, these instructions apply to the Cross-sticks in this chapter and to all other types.

MATERIALS REQUIRED

Few indeed are the materials required for making boomerangs— some wood of the suitable type and a few simple tools.

WOOD FOR BOOMERANGS

Of utmost importance is the wood that goes into a boomerang, yet happily, the proper woods are never scarce in any community. With one exception all types of boomerangs call for soft, light wood, the exception being the curved boomerangs of the Australian type which are made of heavy, hard wood as described in Chapter VI. Wood is the only material from which boomerangs can be made successfully, all efforts to construct them from metal having proved unsatisfactory to date.

The ideal woods are *basswood, tulip (whitewood)*, and select cuts of Number 1 *white pine. Spruce* and *cedar* are usable but less satisfactory. Basswood is the choice above all others; it is usually straight-grained, does not split and splinter as much as most soft woods, is strong for its weight, and very easy to whittle. Furthermore, as compared to many woods, it has the excellent quality of consistency in weight and texture.

There is no better wood than the right piece of white pine, but many pine boards contain so much pitch in the grain as to make them hard to whittle evenly with a jackknife. Moreover, the sticks cut from a board often vary too much in weight. Some lumber yards sell small boards prepared for cabinet work which they label as "select cuts of Number 1 pine." These strips usually come in thicknesses of one-eighth inch and one-fourth inch, these being the exact measurements needed for our purpose. The finest boomerang wood I have ever used has been found by sorting over such strips and selecting light-weight pieces free from pitch. Not all pine comes up to this standard.

Often as perfect a piece of wood as one could desire can be picked up from an old packing box. If it is soft, strong, and straight-grained, it should make a good boomerang. However, much better results will be obtained by securing from a lumber yard some strips of basswood, whitewood, or Number 1 pine in the proper thicknesses. The boomerangs in this chapter call for two thicknesses—one-eighth inch and one-fourth inch. Care should be taken to select clear, straight-grained pieces, free from knots.

Much of the joy of boomerang making will be lost if the wood is so coarse and stubborn that it does not whittle easily and smoothly.

Balsa wood immediately comes to one's mind in connection with boomerangs because of its use in making model airplanes. However, this wood is so light as to be entirely unsatisfactory and has no place whatever in boomerang making. A boomerang made of it is so very light that it cannot be thrown with enough force to carry it.

TOOLS REQUIRED

Boomerang making is essentially jackknife whittling. The only absolutely indispensable implement is a good, sharp, pocketknife. In making boomerangs of ordinary size a three-fourths-inch gouge and a candle will be helpful, and in undertaking very large boomerangs, a wood rasp and a brace and bit will find use. In assembling the boomerangs an assortment of bolts and nuts will be needed, the sizes being stated in connection with the description of each boomerang.

A tube of liquid solder or quick-drying cement will be of assistance in repairing the broken wings of boomerangs and the sticks that split while being whittled.

If one is interested in finished and polished workmanship, a drawknife, plane, and some sandpaper will be useful. Such a quality of workmanship is always commendable in any line of effort, but the typical boomerang is left in a rather rough stage of finish—being a primitive type of instrument, it seems to possess more color and atmosphere if rather crude and irregular in appearance. Such a boomerang performs just as accurately as does a finely polished one. This is a matter of taste, however, and one can finish them to suit his fancy.

The matter of paint is discussed later in the chapter under the heading, "Decorating Boomerangs."

FOUR-WING BOOMERANGS

The four-wing boomerangs are made of two sticks crossed at right angles and bolted in the center, as illustrated in Figure 1. They are frequently referred to simply as Cross-sticks.

When thrown, a light-weight boomerang of this type will cut almost a perfect circle in the air. They move more swiftly than do the Pin-wheel (six-wing) Boomerangs which will be described in the next chapter, but are no less accurate. Large and heavy Cross-sticks frequently travel in a straight line for a long distance, then turn around rather abruptly to the left and return in almost a direct course. However, the characteristic course of flight of this type of boomerang is a circular one.

YOUR FIRST BOOMERANG

One's first attempt at boomerang making should be confined to small sticks. In making the little boomerang here described, we will become familiar with all of the essential techniques in boomerang construction, thus making it possible for us to attempt the larger ones with ease and assured success.

Secure two of the ruler-like sticks used by gasoline stations

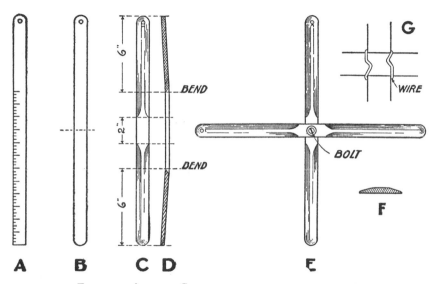

FIGURE 1. A SMALL CROSS-STICK MADE OF GASOLINE STICKS

to measure the amount of gasoline in automobile tanks. These are usually of the right dimensions and made of the right kind of soft wood. Failing here, secure two strips of basswood or other soft wood eighteen inches long, one-and-one-fourth inches wide, and one-eighth inch thick.

Place each stick on the edge of your knife blade and balance it to determine the center of gravity, and mark this point. The center of gravity is used as the center of the stick, rather than the center of measurement. Mark out a two-inch section in the center of each stick as shown in C, Figure 1. With a jackknife, trim off the edges on one side of the stick beyond these lines so that it is roughly convex, as indicated in C, Figure 1, and shown in the diagram of the cross-section, F, in Figure 1. Note that the bottom or back side remains flat, while the top side is roughly rounded

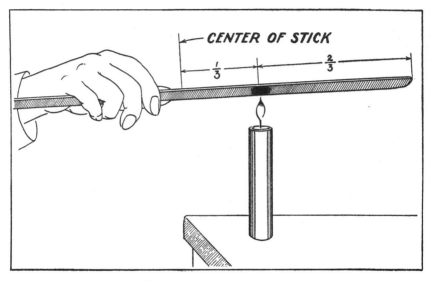

FIGURE 2. HEATING THE STICK PRELIMINARY TO BENDING

off. The edges should be drawn down to a feather line, but the stick remains full width at the center. All this is done with a jackknife and consequently the sticks are somewhat irregularly convex, but so far as efficiency of the boomerang is concerned, this irregularity makes no difference provided the edges are rather uniformly thin. In fact, it is not absolutely essential that the top of the stick be entirely rounded off to a convex shape—it is usually sufficient merely to bevel the edges down to a thin line.

Now each end of the sticks must be given a slight bend upward, that is, toward the beveled or convex side. Figure 1, D, indicates this bend. This is a task for but a moment or two: hold the stick over a lighted candle so that the flame strikes it six inches from the end, as illustrated in Figure 2. When heated hold the stick with the fingers as shown in Figure 3, bend upward slightly, and hold in this position for a few seconds—the curve will then be permanent. A very slight bend is all that is needed—not over a quarter of an inch. Both ends of each stick are bent in the same way; care should be taken in each case to apply the heat at a point the

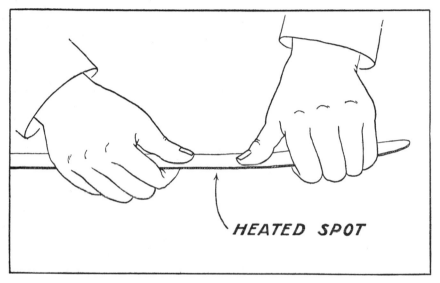

HEATED SPOT

FIGURE 3. BENDING THE HEATED STICK

same distance from the end, and to bend each end about the same amount.

In camp the wood may be bent on a small rock heated in the campfire. When the rock is hot withdraw it from the fire, moisten the stick with the tongue at the point where it is to be bent, then place the moistened spot on the rock and bend to the desired angle. Hold it there for a moment and the curve will be permanently fixed.

To complete the boomerang, place one stick across the other and fasten together, as shown in E, Figure 1—we thus have a flying missile with four wings. Little boomerangs such as this are best held together by wire. Use soft wire that bends easily and run it straight across the stick as shown in G, Figure 1—not diagonally across. When the wire is attached, it may be tightened by grasping it with the pliers and twisting, thus giving it the Z shape indicated in G, Figure 1.

Another very satisfactory method of holding small boomerangs together is to cement them by putting a drop of solder or quick-drying cement at the intersection. This holds the sticks together firmly and permanently, yet adds practically no weight.

Although less acceptable on small boomerangs, the sticks may be bolted together. Use a one-eighth-inch bolt and wing-nut. Wing-nuts are always desirable in that they can be tightened without the use of pliers, and tightening is frequently necessary when the boomerang is in use.

MAKING LARGE FOUR-WING BOOMERANGS

The making of the little Cross-stick Boomerang described in the preceding section will serve merely as an incentive for the fashioning of larger and more colorful ones. Every essential of boomerang construction was involved in making the little boomerang of gasoline sticks, however, and consequently the large Cross-sticks follow the same general pattern and formula.

To make an excellent boomerang of the ideal size for general use, cut two strips of basswood, whitewood, or pine, twenty-four inches long, one-and-one-half inches wide, and one-eighth inch thick. Bevel the edges on the top side throughout the entire length of each stick, bringing them down to a feather line, and round off the top to a roughly convex shape, as described in the preceding section.

The corners at the end of each stick should be trimmed off to the curved shape shown in Figure 4. Note, however, that the end is not cut to a semi-circular shape, but rather, more wood is removed from the left corner of the upper wings than from the right corner—this is true of each wing when it is placed in the upper position. The reason for this is that the left or cut-off side

is the point that cuts into the air as the boomerang sails, and like-
wise it is the point that hits the ground or an obstruction in case
the boomerang should fall; the wing is less inclineed to split when

FIGURE 4.   A TYPICAL CROSS-STICK BOOMERANG

trimmed off in this way.   Do not bevel or thin down the ends
of the sticks—the wood should be left at full thickness here to
insure all possible strength.

Now hold each stick over a candle at a point eight inches from
the end, and when heated, bend it slightly toward the convex or

beveled side, as described in the preceding section (See Figure 3).
Bend each of the four wings in the same way.

*Regardless of the size of the boomerang, the point of bending
is approximately two-thirds of the distance from the end to the
center.* This is clearly illustrated in Figure 2.

Boomerangs two feet or more in length should be bolted rather
than wired. A ⅛- or ³⁄₁₆-inch bolt, ¾ inch long, equipped with
a wing-nut, is proper for all except very large and heavy boomerangs, which latter require a ¼-inch size. The wing-nut should
be placed on the top side of the boomerang, as shown in Figure 4.

Figure 5. Cross-stick Boomerang showing the long bolt used as a handle
in catching

One of the interesting tricks in boomerang throwing involves
the use of a long bolt as illustrated in Figure 5. As the boomerang
returns, the performer catches it by the bolt and the boomerang
thus continues to spin as he holds it in his hand. So fascinating is
this method of catching that it is well to equip all Cross-sticks two
feet or more in length with these long bolts. Using a bolt four
or five inches long, place a nut and washer a half inch from the
end, insert the end through the two blades, place a washer on top,
and hold with a wing-nut. Figure 5 shows the arrangement
clearly. These long bolts should be as light in weight as the size
of the boomerang will permit—a three-sixteenth-inch size is large
enough for all except extremely heavy boomerangs.

No type of nut is satisfactory for use on boomerangs unless it
can be tightened with the fingers, without the use of pliers. Since

the nut tends to loosen in throwing and must be frequently tightened, it is a great convenience if it can be adjusted with the fingers. Two types of nuts meet this requirement, the wing-nut (Figure 7) and the toggle bolt and nut (Figure 6). The latter is particularly desirable on large boomerangs.

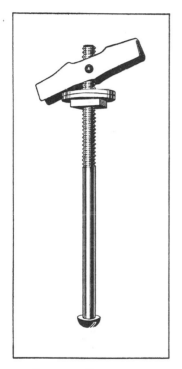

FIGURE 6. TOGGLE BOLT   FIGURE 7. BOLT WITH WING-NUT

When the two sticks are bolted together as in Figure 5, the boomerang is completed and ready to throw.

If the boomerang dives to the ground, or cuts in at you without turning into a horizontal position and floating in easily and gently, it probably is too heavy and needs to be lightened. This is accomplished by gouging out some of the wood on the back or flat sides of the sticks. Use a three-fourths-inch gouge for this. A few chips may be removed as in A, Figure 8, thus lightening the weight slightly, or the entire length of the stick may be gouged out uniformly, making the bottom side concave as shown in B, Figure 8. Most large boomerangs need to be lightened in this way.

The Two Secrets of the Boomerang.—It will be noted that there are two processes in preparing sticks for a boomerang: (1) round-

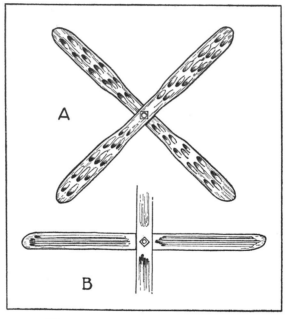

FIGURE 8. WINGS LIGHTENED BY GAUGING OUT THE BACK SIDES

ing off the top side to a roughly convex shape while the bottom remains flat, and (2), bending each wing slightly toward the convex side. *These are the two secrets in boomerang construction.* It is the convex or beveled shape of the top side that is the primary factor in causing the boomerang to return. The bend in the wings causes the boomerang to tend to turn over into a horizontal position and thus to float and sail, staying in the air until it has had time to return. The bend in the wings forms a dihedral angle which helps the boomerang to glide and volplane. An occasional boomerang will be found that will return satisfactorily without the angle or bend in the wings, but this is not often the case—such boomerangs usually tend to dive heavily to the earth before completing the return.

Other "tricks of the trade" in making efficient boomerangs will

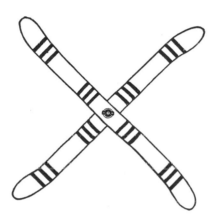

FIGURE 9. CROSS STICK BOOMERANG WITH CURVED WINGS

be discussed in describing the larger Cross-sticks in the pages that follow.

CROSS-STICKS WITH CURVED WINGS

Figure 9 shows a favorite boomerang pattern in which the sticks are slightly curved. These are delightful both in appearance and performance. The dimensions are: twenty-four inches long, one-and-three-fourths inches wide, and one-eighth inch thick. In laying out the wing, cut a strip two-and-one-fourth inches wide and two feet long. Draw the curved wing on this strip, making the wing width one-and-three-fourths inches; then whittle it out. A strip of this width will provide just enough curve to the sides of the wing for a graceful appearance.

In making a boomerang thirty inches long, the width of the wings would be two-and-one-fourth inches and the thickness one-eighth inch. Wings of this width should be drawn on a piece of lumber three inches wide.

The sticks having been cut to the curved shape, the boomerang is completed in the same way as are the straight Cross-sticks, described in the foregoing pages.

JUMBO CROSS-STICKS

The big Cross-stick Boomerang illustrated in Figure 10 is made of boards thirty-six inches long, two-and-one-half inches wide, and one-fourth inch thick. One end is cut down to a smaller size, forming a handle so shaped as to fit the hand, but further than this, the method of construction is just as in the smaller sizes. The back sides of the wings are gouged out to a concave shape throughout the entire length. It is held together by a one-fourth-inch bolt, five inches long.

Such a boomerang is for outdoor use. It goes in a straight line for a long distance, turns abruptly to the left, and returns in almost a straight line. It comes in gently, and can be easily caught by the bolt.

TABLE OF DIMENSIONS FOR CROSS-STICKS WITH WINGS OF UNIFORM WIDTH

The Cross-sticks discussed thus far in this chapter are made with sticks uniform in width throughout the entire length. The table of dimensions presented herewith is for boomerangs of this type. Some Cross-sticks are made with wings that are wider at the ends than at the center—these are discussed in the section following.

Whenever the length of the stick is increased, the width should be increased proportionately. For every six inches that the length is increased the width should be increased about one-half inch.

If we assume as the standard proportion, sticks that are twenty-four inches long and one-and-one-half inches wide, to increase the length to thirty inches would necessitate increasing the width to two inches. This is not a hard and fast rule, since all sorts of odd

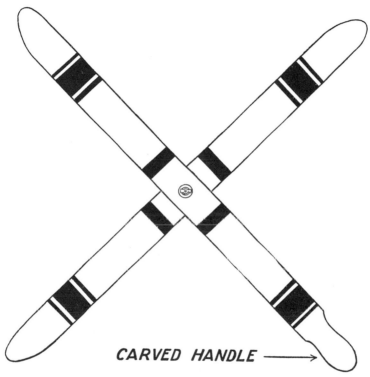

CARVED HANDLE ⟶

FIGURE 10. JUMBO CROSS-STICK

dimensions may be worked out, but these proportions are sure to be satisfactory. If the width is too great for the length, the wings offer too much resistance to the air and the boomerang does not have sufficient carrying power.

The following dimensions are recommended for Cross-sticks with wings of uniform width throughout:

Length	Width	Thickness
18 inches	1 inch	⅛ inch
24 inches	1½ inches	⅛ and ¼ inch
30 inches	2 inches	⅛ and ¼ inch
36 inches	2½ inches	¼ inch

The thickness of the wood is determined by the use to which the boomerang is to be put. Boomerangs one-eighth inch thick are light in weight, cut a small circle, and sail gently into the hands of the thrower. Those of one-fourth-inch stuff carry further and are ideal for outdoor use.

CROSS-STICKS WITH FLARED ENDS

These boomerangs differ from those described thus far in that the wings are wider at the ends than at the center, whereas those

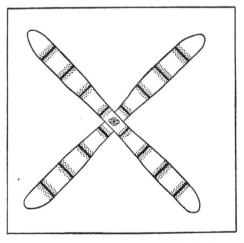

FIGURE 11. THE ENDS OF THE WINGS MAY BE FLARED OUT INSTEAD OF STRAIGHT

described up to this point are made with sticks that are uniform in width throughout. Figures 11 and 12 illustrate flared-end types —they are particularly effective boomerangs. This arrangement lightens the weight and at the same time provides a wide wing at the point where width is needed. Furthermore, with this pattern it is possible to secure the effect that wider wings produce without the danger of too much air resistance which would result if the greater width were continued throughout.

These boomerangs are not only pleasing in their lines but they sail and soar more than those made of uniform sticks, staying in the air longer before returning. From whatever angle we view

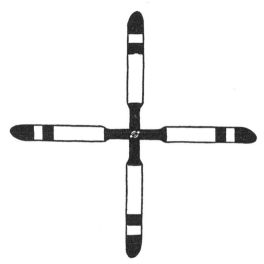

FIGURE 12. ANOTHER TYPE OF FLARED-END BOOMERANG

these flared-end boomerangs they prove to be the most efficient and satisfying type of Cross-stick.

The patterns illustrated in Figures 11 and 12 are twenty-four inches long, one-eighth inch thick, two inches wide at the widest point, and one-and-one-fourth inches at the narrow section. Larger boomerangs may also be made in the same designs provided the same relative proportions are maintained.

CROSS-STICKS WITH TAPERED ENDS

This type of boomerang violates a basic principle of wing construction but it serves well the specialized purpose for which it is designed. The wing of a boomerang should be wider at the end than at the bolt, or at least it should be uniform in width throughout, if typical performance is to be expected. All the boomerangs described thus far are of one or the other of these two types. The wings of this boomerang, however, are wider at the middle and taper down to a narrow end. The boomerang thus has very little wing spread and consequently little capacity to stay in the air. When thrown, it follows the typical course of the Cross-sticks but drops quickly and abruptly to the thrower at

FIGURE 13. CROSS-STICK BOOMERANG WITH TAPERED ENDS

the end of the flight; if it is not caught it drops to the floor at the thrower's feet. It is useful only when one must throw boomerangs under conditions where it would be dangerous if the boomerang were not caught but allowed to float past him. An actor on a very small stage is sometimes in such a situation—a boomerang that returns and circles around him and behind him before coming to his hands might hit the back of the stage; others that cut in directly to him might, if not caught, float past him and out into the audience. If made with tapered ends, he can be sure that the boomerang does not have enough carrying power to go past him. Those boomerangs are certainly not to be recommended for any other purpose, and since few people will ever have to meet such a situation, they are of little use at all. A professional of enough experience to be throwing from the stage will be able to so manipulate ordinary boomerangs that they will not get past him.

To make one of these boomerangs, cut from one-eighth-inch stuff two strips two inches wide and twenty-four inches long. Leaving them two inches wide at the middle, taper them gradually down to a width of one-half inch at the end, as shown in Figure 13. In rounding off the top sides, leave them unbeveled for a distance of two inches at the ends. In other respects the boomerang is completed as any other Cross-stick.

THE CROSS

The Cross-stick Boomerangs described thus far are of the balanced type; that is, the wings are of equal length. The Cross is an unbalanced Cross-stick—one stick is placed across the other at a point near one end, thus giving the boomerang the appearance of a cross as shown in A, Figure 14. The shortest wing is one-third of the length of the stick, and the longest wing two-thirds of the length. The dimensions of the sticks are the same as in four-wing Cross-sticks already described.

The Cross flies with a jerky movement, and while it may be lacking in grace of movement as compared to the others, it returns to the thrower just as efficiently.

Variation.—Another type of Cross which acts in much the same way as the one just described is suggested in B, Figure 14. The long wings are twice the length of the short wings.

TWO-WING CROSS-STICK BOOMERANGS

The boomerangs which we shall now describe differ from the Cross-sticks already discussed in that they have two wings instead of four. These are not particularly effective boomerangs. Certainly they do not compare at all favorably with the balanced

Cross-sticks in flying qualities, but nevertheless, they are interesting novelties that will appeal to all who find fascination in boomerangs. Once a person is exposed to boomerang making, he is seldom satisfied until he has made every conceivable type that can be caused to come back to him. These two-wingers will return

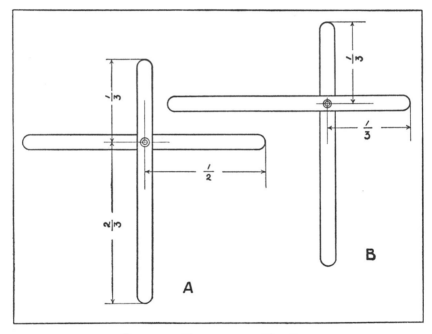

FIGURE 14. TWO STYLES OF THE CROSS

even though they do not possess the precision and the soaring qualities of the other models.

Boomerangs of these types should be thrown higher in the air than is usually the case (see Chapter VIII "How to Throw Boomerangs"). They go straight for a short distance, turn to the left and rise in the air, hesitate, and then glide and volplane down to the thrower. They cannot be caused to make a circular flight as do the regular Cross-sticks.

THE SQUARE

This boomerang consists of two blades arranged so as to form a right angle, as illustrated in A, Figure 15. It is in fact one-half of a four-wing Cross-stick Boomerang. The dimensions are the same as in the Cross-sticks except that the wings are one-half the length plus two inches for overlapping and bolting. Take any

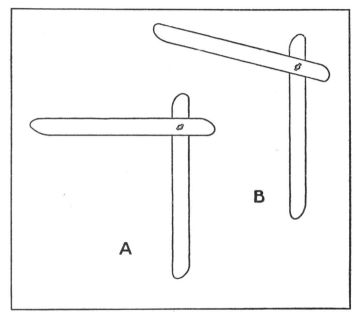

FIGURE 15. THE SQUARE

of the suggested dimensions for Cross-sticks on page 26 and shorten the sticks as here suggested. The sticks are beveled as usual and bent just as in making the regular Cross-sticks, but only in one place, that point being two-thirds the distance from the end to the bolt (see Figure 2). Arrange the wings so as to form a right angle and fasten with a lightweight bolt and wing-nut.

Variation.—A much more effective boomerang than the one

just described is illustrated in B, Figure 15. The dimensions and proportions are the same except that the blades are longer, extending three-and-one-half inches beyond the bolt. The wings are adjusted so as to form an angle slightly greater than a right angle. These boomerangs are somewhat similar in appearance to the curved Australian type, but they are very different in action.

THE ICE-TONGS

In this pattern the sticks are cut as shown in Figure 16. They are beveled and bent as in making the Square described above.

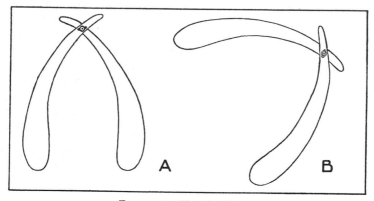

FIGURE 16. THE ICE-TONGS

When the wings are set at a small angle as shown in A, the arrangement takes on the appearance of ice-tongs, hence the name. The wings may also be set at a larger angle as shown in B. These novelties are not particularly efficient as boomerangs but are interesting nevertheless.

THE RAZOR

The sticks are cut to represent a razor, as illustrated in Figure 17. The following dimensions are recommended, the width applying to the widest points:

Width	Length	Thickness
1 ¾ inches	15 inches	⅛ inch
2 ⅛ inches	19 inches	⅛ to ¼ inch
2 ½ inches	22 inches	¼ inch

The sticks are whittled to the usual convex shape on top, and are bent just as in making the Square described on page 31. They

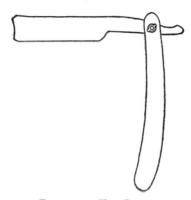

FIGURE 17. THE RAZOR

should be adjusted so as to form an angle a little larger than a right angle.

DECORATING BOOMERANGS

No boomerang is complete until it is painted. Coloring adds immeasurably to the eye appeal both when the boomerang is in the hand and in the air. Properly ornamented with stripes across the wings, the whirling Cross-sticks create vivid circles of color which are strikingly effective.

PAINTING

Do not use ordinary paint or enamel on the boomerangs—it is heavy and creates too thick a layer. Secure some aluminum paint powder together with the oil in which to mix it. The silver effect produced by the aluminum is particularly effective and easily visible against almost any background. It dries quickly and adds practically nothing to the weight. Mix together a little powder and oil, making a thin paste, and paint with it.

Also, secure from a paint store a little paint powder or dry pigment in vivid yellow, ultramarine, and fire red. This powder is very inexpensive. Fire red is the only red powder that is effective on boomerangs—it is very bright and vivid as contrasted to the dullness of the usual red powders. It can be obtained from the larger paint stores.

Put a little powder of each color in containers, add a dash of aluminum powder to each, and mix with the aluminum oil until a thin paste is formed. The adding of the aluminum contributes a luster or frosted appearance, and helps the color to stand out in artificial light. Paint prepared in this way gives a theatrical appearance to the boomerangs.

The background color of a boomerang should be aluminum, yellow, or fire red. The blue is suitable only for striping. The boomerang must first of all be clearly visible in any kind of light, and this is best accomplished by the use of lighter shades for the background color, set off with stripes of darker colors. Paint the boomerang in aluminum and add crosswise stripes to the wings in blue, yellow, or red. Or paint it in yellow or red with stripes of aluminum, edged with thin lines of blue.

Each wing of the boomerang should be painted in the same way or else the design will be lost when the missile is spinning. There are many effective ways of striping the wings. Excellent designs for Cross-sticks are shown in Figures 4, 9, 10, 11, and 12. The possibilities are limitless here and each craftsman will be able to work out many unique and original effects of his own.

ILLUMINATING BOOMERANGS

The attaching of electric lights to boomerangs gives a unique and brilliant effect, producing the appearance of whirling circles of fire. Secure tiny flash light bulbs and batteries. Attach the battery with wire as near as possible to the bolt at the center. Bore a hole through the wing midway between the bolt and the end, set the bulb well down in this hole so that it is protected from damage, and run the wire to the battery. Two bulbs are enough to produce the circle of light as the boomerang spins—they should be placed on wings directly opposite each other, that is, one on each wing of the same stick, equally distant from the bolt. If two circles of light are desired, add two additional bulbs on the other set of wings, placed at a different distance from the ends than the first set.

COLORED STREAMERS

If used at all, colored streamers and trailing attachments must be handled cautiously and very conservatively. Nothing in the way of trailers can be attached to a boomerang without impeding its flight and cramping its style more than one would think. Certainly nothing of this sort, however light it may be, can be attached to the ends of the wings without crippling the boomerang rather completely. It is possible to attach slender and very light colored ribbons to the bolt on the under side of a boomerang provided it is sturdy and buoyant enough to offset the drag. It is not so much the weight of the ribbons that hampers the flight as it is the wind resistance against them as the boomerang flies through the air. Air currents are doubly detrimental when streamers are used, often causing the boomerang to become exceedingly erratic. To be effective, the ribbons should be at least two feet long.

CHAPTER III

Pin-wheel Boomerangs—How to Make Them

If the Cross-stick can be glorified as the most perfect type of boomerang, certainly to the Pin-wheel goes the credit of being the most perfect Cross-stick. That puts it far and away at the head of the list of all the boomerangs. Made of three sticks bolted together in the center, it is a six-wing boomerang such as is illustrated in Figure 18.

Like huge pin-wheels these boomerangs go humming around their circuits. Having more blades than the four-wingers, they are more steady, more exact, more dependable—they find their way back home with the utmost of precision. When one throws a Pin-wheel he knows that it is going to travel exactly the same course in exactly the same way that it did last time—this owing to its many wings, for if one of them should become warped or twisted, there are five others to offset any damaging effects.

One knows, too, when he hurls a Pin-wheel, that it will not come back bluntly, hopping and jerking, abruptly thrusting itself at him. The Pin-wheel has too much refinement for that—by nature it is too gentle, too sensitive. It soars and sails—slowly, gracefully, aesthetically—and floats in to you gently, settling softly into your hands. Herein rests its chief superiority over those with fewer wings, which latter zip about swiftly, cut a close circle, and often dive in to you so suddenly fast that you scarcely expect them. A good Pin-wheel often seems reluctant to come home to you at all—having drifted lightly back it turns and floats away again in another little circle before finally settling into your hands. Indeed, there are some that make three trips away before coming home at last.

Both in precision of returning and in beauty of action, there is no boomerang so satisfying to handle as the Pin-wheel.

This chapter will describe first the making of the six-wingers—the typical Pin-wheels—and then it will take up the three-wing boomerangs.

MAKING PIN-WHEELS

The sticks for the Pin-wheels are prepared precisely as for the simple Cross-sticks described in the preceding chapter. It takes three of them to make a Pin-wheel rather than two, but there is

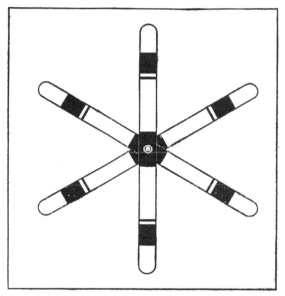

FIGURE 18. A TYPICAL PIN-WHEEL WITH WINGS OF UNIFORM WIDTH

no difference in their preparation. In fact, if one of the sticks of a Pin-wheel should break, we can usually arrange the two remaining sticks at right angles to each other and thus convert the wreckage into a serviceable Cross-stick of the four-wing type.

Since every one should try his luck at making one of the four-wingers discussed in the last chapter before attempting a Pin-wheel, the process of preparing the sticks should by now be familiar to all, but let us briefly reiterate the essentials just as a precaution against any possible oversight: Bevel the edges with a jackknife on the top side of each stick throughout the entire length (see the discussion on page 16 and the illustration in Figure 1). Do not bevel the ends but rather leave them at their full thickness, doing nothing more than to trim off the corners to the shape shown in Figure 18. Bend each wing as illustrated in Figures 2 and 3, and described on page 17. Fasten the three sticks together at their centers with a three-sixteenth-inch bolt, four inches long, and a wing-nut. The result is a balanced boomerang of the cross-stick type which possesses six wings uniform in size and shape, as depicted in Figure 18.

Like the simple Cross-sticks, the Pin-wheels may be made with sticks either uniform in width throughout, or with flared ends. The flared-end type will be discussed in a separate section presently.

Of the Pin-wheels with wings uniform in width, so many are of one of two sizes that these sizes may be regarded as standard:

The first of these is pictured in Figure 18—a delightful little Pin-wheel, excellent in performance, easy to handle, and always dependable. The sticks are twenty-four inches long, one-and-one-half inches wide, and one-eighth inch thick. This is a popular Pin-wheel always.

The second of the standard sizes is more impressive because of its greater size, and is unquestionably more brilliant in action. It is made of sticks of the same width as the first size, but thirty inches long. The grace and ease with which this six-winger sails makes it a favorite at any time and in any place. Big as it is, it cuts such a small circle that it can be thrown in any small gymnasium or hall. The sticks are one-and-one-half inches wide, thirty inches long, and one-eighth inch thick. It is obvious that sticks of these dimensions would not make a very satisfactory four-wing Cross-stick in that they are too long and slender, but since the Pin-wheels have six wings, the sticks used in them may be narrower in relation to the length than is ordinarily the case. Too wide a wing spread will prove unsatisfactory in a Pin-wheel.

JUMBO PIN-WHEELS

Huge size is always an impressive factor in the Pin-wheels. An oversized wheel can be so constructed as to belie completely its weight and its flying distance. Over-large in all dimensions, such a wheel looks forbidding indeed when picked for throwing indoors, yet in reality it may be very light and may cut as small a circle as any young-sized Pin-wheel. These jumbo wheels thrill by their very hugeness when they are held up preparatory to throwing, yet they prove to be as harmless as their tiniest brothers.

Out of one-fourth-inch stuff, cut three boards forty-two inches long, and two-and-one-fourth inches wide. Whittle off the top sides to a convex shape and bend the ends as usual in preparing sticks for a Cross-stick Boomerang. Shape the end of one of the sticks into a handle by cutting it down to fit the hand, as illustrated in Figure 19. Using a one-inch gouge, remove as much wood as possible from the back side of each stick, gouging it out to a concave shape. The gouging should be carefully done so that the stick becomes a thin shell, not more than one-eighth inch through. Bolt the boards together at their center in the usual fashion, using a one-fourth-inch bolt, five inches long.

It is sometimes difficult to fasten these over-large wheels together with a bolt tightly enough so that the wings will not move when thrown. If this occurs, the boomerang either collapses and falls, or nose-dives abruptly to the ground. To guard against such a happening, a very small hole may be bored through each wing about five inches from the bolt and a wire run through them as illustrated in Figure 19. This will preclude any chance of the blades moving enough to influence the flight. Such precautions should always be taken when large wheels are thrown indoors.

Still another style of Jumbo Pin-wheels may be made by flaring out the ends of the wings. Use boards forty-five inches long, three inches wide, and one-fourth inch thick. Divide each board into three fifteen-inch sections; leave the two end sections at their full width of three inches, but pull the center section in to two inches in width. Gouge out the back side and complete the boards as usual.

HEAVY PIN-WHEELS FOR OUTDOOR USE

All the Pin-wheels described thus far in this chapter may be used either indoors or out. A long-flying Pin-wheel for strictly

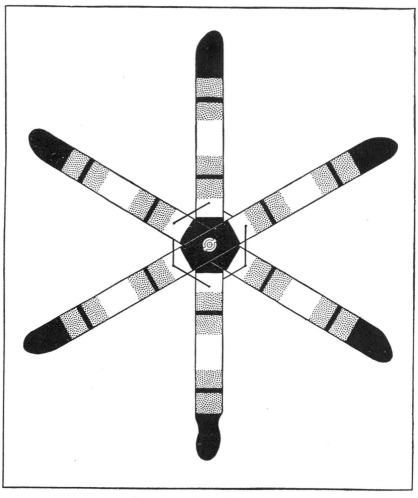

FIGURE 19. JUMBO PIN-WHEEL

outdoor use need not necessarily be large in size but rather must be heavier than the average in weight.

An excellent far-flying Pin-wheel of the larger size may be made from sticks thirty-six inches long, one-and-three-fourths inches wide, and one-fourth inch thick. Leave the sticks unbeveled and at their full thickness for a space of four inches at the center, and whittle them off to the usual convex shape beyond this area. Such boomerangs are impressive in the great distance that they travel, but they do not sail and float so gracefully as the lighter ones.

An equally good long-flyer of the smaller size calls for sticks twenty-four inches long, one-and-one-half inches wide, and three-sixteenths to one-fourth inch thick. Those of one-fourth-inch stuff may be gouged out a little on the back side if it is necessary to lighten the weight for good performance.

TABLE OF DIMENSIONS FOR PIN-WHEELS
WITH WINGS OF UNIFORM WIDTH

Many odd and lawless sizes and shapes are possible in Pin-wheels, but the dimensions set forth in the following table are sure to give satisfactory results. By the time these sizes have all been successfully made, the creative urge that grips all boomerang enthusiasts, once they are really bitten by the bug, will lead to the quick discovery of other proportions.

Length	Width	Thickness.
24 inches	1½ inches	⅛ inch
24 inches	1⅛ inches	⅛ inch
30 or 32 inches	1½ inches	⅛ inch
36 inches	1¾ inches	¼ inch
36 inches	2 inches	⅛ inch
24 inches	1¾ inches	3⁄16 inch
30 inches	1¾ inches	¼ inch
42 inches	2¼ inches	¼ inch
45 inches	3 inches	¼ inch

PIN-WHEELS WITH FLARED ENDS

There could scarcely be an argument on the point that wings with flared ends produce better Pin-wheels than those that are uniform in width throughout. Such wheels take to the air with greater ease, conduct themselves with more charm and gentle grace, yet maintain withal a buoyant and enlivened spirit, all of which results from the fact that there is greater width of wings where width is needed for flying, and less where width serves only as dead weight to be carried.

While no indictment of the excellent Pin-wheels described thus

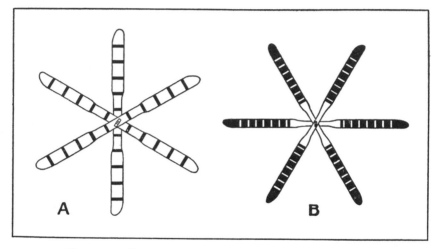

FIGURE 20. PIN-WHEELS WITH WINGS FLARED OUT AT THE ENDS

far in this chapter is meant, the boomerang maker will do well to fix in his mind as the type-perfect model, the wing with ends flared out wider than the center.

Figure 20 shows two characteristic outlines for such widened ends. In A the center area is pulled in gently and with curving lines, whereas in B, the center is cut away more deeply and sharply.

In making such a boomerang, divide the stick into three equal sections, the center third to be cut away to a narrower width. Any of the dimensions for Pin-wheels referred to earlier in this chapter may be used as a basis for the flared-end type. For any one of the dimensions given, merely narrow down the center section a quarter or half inch, leaving the ends at the stated width. For example, if the dimensions of the stick are twenty-four inches long by one-and-one-half inches wide, cut down the center to one inch, leaving the stick an inch and a half wide at the ends.

The boomerang used as a model for the drawing shown in A, Figure 20, is a large and very effective Pin-wheel, measuring three feet across. For a distance of one foot at each end, the sticks are

FIGURE 21. AN EXCELLENT, LONG-FLYING PIN-WHEEL

two inches wide, and the one-foot center area is one-and-one-half inches wide. The thickness is one-eighth inch full.

The model for B, Figure 20, is a twenty-four inch boomerang made of slender sticks. They are one-and-one-eighth inches wide at the ends, seven-eighth inches wide at the center. It floats lightly and beautifully.

Figure 21 shows one of the very finest Pin-wheels ever encountered in the experience of this writer. It is a long-traveling wheel with remarkable soaring qualities. When used out-of-doors and thrown with much force, it will travel far, rise high in the air, and will often make three circles before returning. The dimensions of the sticks are: twenty-four inches in length, two inches in width at the ends, one-and-one-half inches at the center, and three-sixteenth inch in thickness. The three-sixteenths thickness gives it a greater weight than most Pin-wheels and consequently greater carrying power. Note the curving lines of the wings.

Figure 22 shows a colorful wheel of good flying qualities which

FIGURE 22. A PICTURESQUE PIN-WHEEL

has the ends shaped and painted to represent birds' heads. The sticks are thirty-two inches long, one-eighth inch thick, and one-and-one-half inches wide except for the heads which are two-and-one-fourth inches wide.

THREE-WING BOOMERANGS

These boomerangs with three wings have one-half the wing support of the Pin-wheels. Their course is a circular one similar to that of the four-wing Cross-sticks and the Pin-wheels, but since they have less wing support than either of these, they move more swiftly, often having a jumpy action, and by comparison are lacking in graceful sailing qualities.

MAKING THREE-WING BOOMERANGS

The sticks are prepared just as in the case of the ordinary Cross-sticks and Pin-wheels—whittled to a convex shape on top and bent

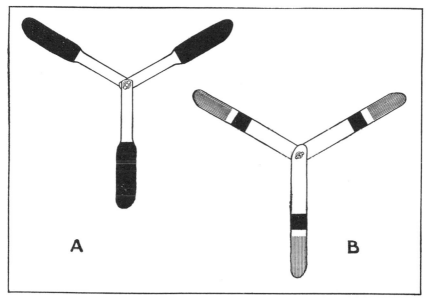

FIGURE 23.   THREE-WING BOOMERANGS

—but since each stick constitutes one wing rather than two, it is bent in one place only, that point being two-thirds of the distance from the end to the bolt.   When bolted together, the sticks are arranged equidistant from one another, as shown in Figure 23.

These three-wingers may be made either with wings flared out at the ends as in A, Figure 23, or with wings uniform in width throughout, as in B.   In boomerangs of this type, the flared ends

are by all odds preferable. Not having much wing support anyway, the widened ends are in many cases essential to good performance. Without them the boomerang zips about with a
hopping motion, making it difficult for the thrower to follow
its course, and it cuts in to him too swiftly to be handled with
ease. Such boomerangs are often called "hoppers."

A good wing length for a three-winger with flared ends is sixteen inches. Make the wings two-and-one-half inches wide at

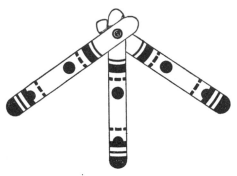

FIGURE 24. THE FAN

the wide area and pull them down to one-and-one-half inches at
the center area. The wide section should extend about one half
the length of the wing. The thickness is one-eighth inch.

For "hoppers," those with wings of uniform width such as the
illustration shows in B, Figure 23, the following dimensions are
recommended:

Width	Length	Thickness
1½ inches	12 inches	⅛ inch
2 inches	16 inches	⅛ inch
2¼ inches	18 inches	⅛ inch

FANS

The Fan, illustrated in Figure 24, is in fact one-half of a Pin-
wheel: if three adjacent wings of a Pin-wheel are cut off near the
bolt, we have the Fan. The three-wing boomerangs in Figure 23

may be converted into Fans by loosening the bolt and adjusting the wings as shown in Figure 24.

While the fans can be thrown so that their flight will be similar to that of the Pin-wheels, their most characteristic action is to sail high up in the air, turn into a horizontal position, and then glide and volplane back in almost the same plane to the thrower.

CHAPTER IV

Boomabirds—How to Make Them

"All birds come home to roost"—and so it is with the Booma-bird. Like the homer pigeon, it invariably finds its way back to its master. Of all the boomerangs, it is the most picturesque, whether held in the hand or flying over the tree tops. As these bird-shaped boomerangs gracefully sail and soar overhead, they look for all the world like living, gliding birds.

It was on one of those calm, peaceful evenings in June when not a breath of air was stirring that I picked up some Boomabirds and went over in the shadow of a great football stadium. Never did the boomers soar so beautifully—the quiet evening was made to order for their liking. There was but one annoyance—a flock of swallows circled and circled overhead, making it difficult to keep one's eye on the Boomabirds. Suddenly as a Boomabird settled into my hands, a swallow swooped down and almost hit me! Then the explanation of it all abruptly dawned—*the swallows were being attracted by the Boomabirds. So realistic were these bird-shaped sticks that they fooled even the birds themselves.*

Since then the birds have joined company with the Boomabirds on many an occasion. They seem to feel that since the Booma-birds keep circling around one spot, there must be some choice attraction there of no ordinary interest to birds and so they proceed to investigate for themselves.

At first blush it would seem that when so much attention is given to the shape of a boomerang as to make it resemble a bird, something in the way of efficiency must be sacrificed. Happily, however, this is far from being the case. These birds have a flying style all their own, but they find their way back to your waiting hands with such unerring propriety, and withal so gracefully, that they must be listed high up in the social register of the boomerang élite. They hold their own among all others in finished etiquette and they excel outstandingly in glamour and personal attractiveness. Their very life-like, bird-like appearance as they glide about enhances their grip on the imagination.

When thrown, the Boomabird goes straight forward for some distance, swings to the left and rises high in the air, then circles over the head of the thrower and around in front of him, reverses its direction and settles gently into his hands. The average Boomabird makes two circles before reaching the thrower, and occasionally one is found that will make three circles.

The author had a small Boomabird at one time that, when thrown from a theater stage, would sail up above the balcony and start back, making a complete circle over the audience before floating up to the stage. As it approached him, the thrower would start off the stage to the right and the bird would follow him out the wings. Such a Boomabird is one in a million, of course, but the Boomabirds in general are noted for their odd and intricate tricks. Their lawless habits make it impossible to predict their conduct until each has been duly subjected to a detailed case study.

A Boomabird is in fact a four-wing Cross-stick Boomerang. It differs from those described thus far in that one of the sticks, the body stick, is much wider and heavier than the stick used for the wings.

MAKING THE BOOMABIRD

Boomabirds require the same light, soft wood described for the Cross-sticks in Chapter II: basswood, tulip (white-wood), or Number 1 white pine.

Figure 26 illustrates the best shape for the body of the bird. The head is usually the same width as the body, but may be slightly narrower. Figure 27 shows another type of body that some may prefer.

An excellent size for all-around use is diagrammed in Figure 25. Cut a board from one-eighth-inch stuff, twenty-two inches long, and three-and-one-fourth inches wide. Draw the outline of the bird on it, as illustrated in A. Note that the body is divided into three parts; ten inches should be allowed for the body proper, and six inches each for the head and tail. Whittle out the shape of the bird with a jackknife (B) and then bevel the edges on the top side, rounding them off to a smooth curve as shown in the diagram of the cross section in F. (See page 16 for a more detailed description of the beveling.) This beveling is carried

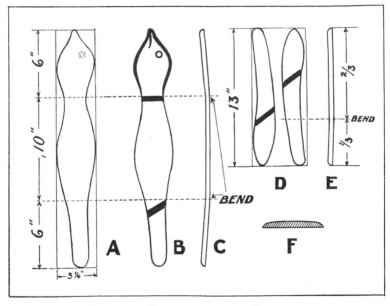

FIGURE 25. PLAN FOR MAKING A BOOMABIRD

all around the body with the exception of the point of the beak and the tip of the tail, which points are left at full thickness.

Now heat the wood at the points indicated in C, Figure 25, by holding over a candle, and give the head and tail a slight bend upward, that is, toward the beveled side. (See Figures 2 and 3, and page 17 for a detailed description of the bending process.)

To determine the point on the body where the hole is to be bored for the wings, place the body on a knife blade until it balances—this point will be an inch or so nearer the head rather than at the exact center of the body.

The wings may be made either of two pieces, as shown in Figure 26, or of one straight piece as shown in Figure 27. The use of two separate wings makes a slightly more attractive Boomabird, but not necessarily a more efficient one.

Let us describe the making of the two separate wings first. As shown in D, Figure 25, each wing is thirteen inches long;

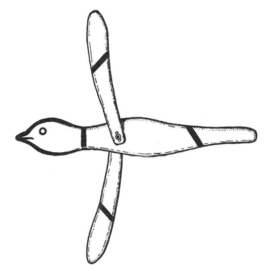

FIGURE 26. A TYPICAL BOOMABIRD WITH SEPARATE WINGS

they are one-and-five-eighths inches wide at the widest point, and one inch at the narrow point. The wings are cut to the curved shape illustrated. Bevel them as in making the body, and bend them at the point illustrated in E—two-thirds of the distance from the end to the bolt.

Bolt the wings in place as shown in Figure 26, using either a short bolt and wing-nut, or a long bolt with which to catch the bird, as described on page 20. Note that the wing on the belly side is placed on the bottom side of the body, and the other on the top side. Tighten the nut very securely.

The wings should first be adjusted so that they are at right

angles to the center line of the body. If the bird does not perform perfectly with wings in this position, turn them at an
angle, nearer the head as shown in Figure 26. Continue to ad-
just the wings in this way until the bird flies in the way that
you want it to.

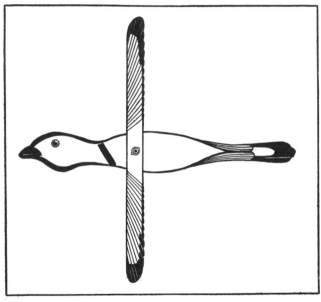

FIGURE 27. ANOTHER STYLE OF BOOMABIRD SHOWING WINGS MADE OF ONE PIECE

If a single stick is to be used for the wings, it should be twenty-
four inches long and one-and-one-half inches wide. Bevel the
edges, and round off the ends as shown in Figure 27. Bend the
ends as before, at a point two-thirds of the distance from the end
to the bolt. Place this stick across the top side of the body and
bolt securely.

Another type of wing is illustrated in Figure 29. It is one
piece of wood cut so as to give the wings a slight curve.

If the Boomabird proves to be too heavy to float well, gouge
out the back side of the body throughout the entire length, as

described on page 21. It will not be necessary to gouge out the wings in Boomabirds of this size.

TABLE OF DIMENSIONS FOR BOOMABIRDS

Boomabirds can be made in a wide variety of sizes. In the following table of dimensions the thickness of the stick is the same for both body and wings. The width of the wing in each case refers to the width at the widest part.

| | Body | | | Wings | |
	Length	Width	Thickness	Length	Width
1.	15 inches	2½ inches	⅛ inch	17 inches	1¼ inches
2.	18 inches	3 inches	⅛ inch	21 inches	1½ inches
3.	22 inches	3¼ inches	⅛ inch	24 inches	1⅝ inches
4.	30 inches	4 inches	¼ inch	32 inches	2 inches
5.	36 inches	4½ inches	¼ inch	39 inches	2¼ inches

Number 3, the twenty-two-inch length, is the one described in detail in the preceding pages. It and Number 2, the eighteen-inch length, are the two best sizes for all-around use. Number 2 is a delightful little Boomer—it usually has a full bag of tricks which it displays generously when flying. Since no two Boomabirds act in exactly the same way, one can make several of these birds, all of the same dimensions, and be able to display a different brand of stunt flying with each.

Number 4, the big thirty-inch bird made of one-fourth inch stuff, is steady, sure, always dependable in its flight, either outdoors or indoors. It flies slowly and deliberately, and returns with studied precision. The back side of both the body and the wings of this bird should be gouged out to a concave shape throughout the entire length.

All Boomabirds twenty-two inches or more in length should be equipped with bolts four inches long. It is easy to catch the birds by these bolts, but difficult without them. Furthermore, the spinning of the bird in the hand when caught by the bolt adds a colorful touch.

THUNDERBIRDS

The Redman's picturesque thunderbird, glamorous always in its eye appeal, suggests the outline for the Boomabird shown in Figure 28. The dimensions and the manner of making are the same as in the Boomabirds described in the preceding pages. A few liberties may be taken with the recommended dimensions

FIGURE 28. THE THUNDERBIRD—A COLORFUL BOOMABIRD IN THE INDIAN STYLE

in order to gain enough width of body to secure the typical lines of a thunderbird, but if the body is made too wide the bird will coast about playfully in the air without tending to its business of coming home.

The design of the thunderbird is painted on the body and wings. Here, as in the handling of all things Indian, we do better to copy in accurate detail an original Indian thunderbird design. The present-day artist thinks to improve on the Redman's ideas of design, but this invariably to his chagrin, for the Indian's simple figures "make medicine" whereas the white man's "improvement" is often nothing more than so many lines on a page. The Red Artist is supreme in his own field and the wise will copy and not attempt to alter or perfect.

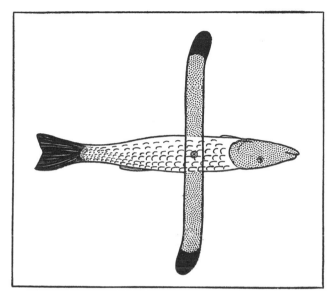

FIGURE 29. A FLYING FISH

FLYING FISH

The flying fish illustrated in Figure 29 is made exactly like the Boomabirds. Any of the dimensions listed for Boomabirds will apply with minor alterations. Here we must think in terms of a long slender type of fish—a northern pike or musky will make a better model than a bass or wall-eye. Use your privilege of artistic license and give the fish long sweeping lines.

BOOMAPLANE

Boomerangs and airplanes somehow seem to go together. People interested in boomerangs are often interested in model airplanes, and aviation enthusiasts usually take to the boomerang hobby wholeheartedly once they are exposed to it. Certainly an airplane suggests a most appropriate shape for a boomerang.

Such a boomerang is represented in the outline of the Booma-plane in Figure 30. Numberless variations in shape are possible, following the lines of any type of airplane. The only requirement is that the relative proportions of body and wings be ap-

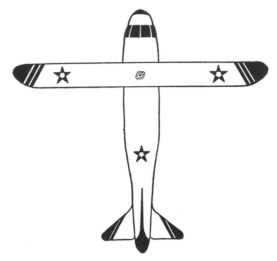

FIGURE 30. THE BOOMAPLANE

proximately the same as in the case of the Boomabirds, and consequently they should be made in about the same dimensions as specified for Boomabirds on page 53. The dimensions may force us to distort the shape of a selected airplane somewhat, but not so much so that the model will not be recognizable.

DECORATING BOOMABIRDS

The preparation of paint for use on boomerangs has already been described on page 34. Remember that light colors are desirable for the main body of the Boomabird in order that it be visible in all kinds of light and against all types of backgrounds —use aluminum, yellow, or fire red. The lines should be in contrasting colors, perferably blue or black. If the blue is used for the body of the bird, ample splotches of aluminum or yellow should be added. Whatever colors are used, a little aluminum powder should be mixed with it to provide luster.

The body of the bird and the wings should be painted in solid color and then outlined in contrasting color. It is wise to paint a stripe across the wings, tail and neck, these serving to create circles as the birds whirl through the air. Spots of contrasting color may also be added to the body. Excellent designs for Boomabirds are shown in Figures 26 and 27.

CHAPTER V

Tumblesticks—How to Make and Throw Them

A Tumblestick is a curious affair—it is essentially a straight stick of wood which when thrown into the air will return to your hands. Since a Tumblestick will come back to the thrower it properly belongs in the category of boomerangs, but certainly it is the most unique and little-known of all the gadgets that can be so classified. Consequently it is the most intriguing type, both to the spectators and the thrower. The Tumblestick never fails to mystify. When one sees a large and elaborate contrivance such as a Pin-wheel, he assumes that there is a mechanism involved that he does not understand which brings it back, but when a straight stick is caused to return the stunt seems so incredible at first exposure as to be baffling indeed.

Every one has seen children throw a slender strip of wood in the air so that it makes a humming sound—whir-r-r-r sticks, the boys call them. These whir-r-r-r sticks resemble a Tumblestick both in appearance and sound, and in fact one is occasionally seen that turns over in the air and makes a feeble effort at returning. Trimmed up perfectly by an expert such a whir-r-r-r stick might conceivably be converted into a dependable Tumblestick.

Because of its very simplicity there is no boomerang so easy to whittle out as the Tumblestick, yet there is no type of boomerang which offers so many difficulties when it comes to producing one of perfect performance—you may make a dozen Tumblesticks exactly alike in size and shape, all out of the same wood, yet find only one that will return accurately and consistently. This delicacy and temperamental quality, however, adds interest to the Tumblestick. One can whittle them out so easily that he does not become discouraged if his first few attempts are unsuccessful, and when at last he finds one that does work, joy is unbounded! A good Tumblestick is precious indeed and is always carefully guarded! The long search for the right stick may convince one that Tumblesticks are too frail and temperamental to bother with, but this much is beyond question: once a good Tumblestick is discovered, it can be depended upon to do its stunt of returning to your hand every time it is expertly thrown. What more could be asked of any boomerang?

MAKING THE TUMBLESTICKS

Look around until you find a light, straight-grained piece of one-eighth-inch basswood or whitewood. Pieces of the same kind of wood frequently vary considerably in weight. The need here is for as light and as evenly balanced a piece as can be found in the kinds of wood mentioned.

Cut from the one-eighth-inch stuff a strip twenty-four inches long and one-and-one-half inches wide. Bevel the edges and round off the top side to a convex shape just as in making the wings for the Cross-stick Boomerangs described in Chapter II. The bottom side remains flat as usual. Do not round off the ends as in making the boomerangs previously described, but rather leave them square as illustrated in Figure 31.

Now throw the Tumblestick as described later in this chapter, to determine whether or not it will return. If after a thorough trial, it does not come back so that you can catch it without stepping, bend each end of the stick slightly toward the beveled side as is done in making the wings of Cross-stick Boomerangs (see Figure 3, also the discussion on page 17. Each end of the stick should be bent in this way, at a point eight inches from the end. Now throw the Tumblestick again, and if it still does not work, try rounding off the square corners at the ends. Most Tumble-

FIGURE 31. THE
TUMBLESTICK

sticks, however, work better with square ends. If the Tumble-stick still does not work, try lightening it by shaving off more wood on the top side, making the stick a little thinner. If this fails, use the stick as part of a Cross-stick Boomerang and make another Tumblestick.

It should not be assumed from all these various suggestions that there is anything particularly forbidding and discouragingly difficult about making a successful Tumblestick. They can be whittled out in five minutes. If failure results when the stick is carefully prepared as described, the difficulty probably rests in the wood—it may be too heavy or unevenly balanced. It may very easily happen that you will be fortunate enough to get a first-class Tumblestick in your first attempt.

It will be noted that the Tumblestick is prepared exactly as are the sticks for a twenty-four-inch Cross-stick. It is always possible, therefore, to use those Tumblesticks that fail to function accurately as parts of Cross-stick Boomerangs. This works the other way around, too—if you take apart your Cross-sticks and try throwing the sticks as Tumblesticks you may discover one that is so perfect in performance as a Tumblestick that you will want to save it for that purpose and replace it with another piece in the Cross-stick whence it came.

While Tumblesticks may be made in larger and smaller sizes, the size suggested above will produce the best results. It is difficult to throw a larger size in that the greater width prevents the stick from whirling in the air rapidly enough. In making other sizes, use the dimensions recommended for Cross-stick Boomerangs on page 26.

THROWING THE TUMBLESTICKS

The methods of throwing boomerangs are discussed in Chapter VIII, but since Tumblesticks require a technique all their own, we shall describe the specialized process of throwing them here.

Grasp the end of the Tumblestick in the right hand between the thumb and forefinger, with the beveled or convex side toward

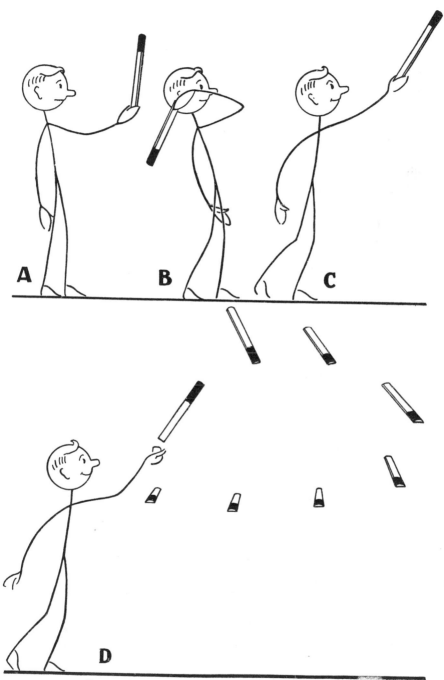

FIGURE 32. How the Tumblestick is thrown

you. Hold the stick straight up by the side of your head as in A, Figure 32. Swing it straight back so that the end is behind your shoulder (B), and then throw it forward and upward over the level of your head. In releasing it there are two important things to remember: *Turn your wrist sharply downward* allowing the stick to rotate forward in your fingers so that it leans a little forward of perpendicular at the moment of release (see C, Figure 32)—this causes it to turn over in the air and start back. At the same time that this is done, give the stick a side-wise twist so that it whirs and hums in the air—this is accomplished by twisting the wrist sharply inward (to your left), thus turning the hand over so that when the stick is released the palm of the hand is up as in D. Do not use much muscle—it is all done with the forearm and wrist.

The Tumblestick will hum and whir, go up in the air a few feet, turn over, settle into a horizontal position, and float back into your hand. The course that the stick follows is illustrated in D, Figure 32.

In catching it, let it float in close to you, and then grab it with a sudden thrust of your right hand. It is easy to let the Tumblestick get away from you in attempting to catch it. Keep your eyes glued on it, and when certain of its location, snatch for it quickly and decisively. The stick is so light that there is positively no danger of injury to the fingers.

A little testing will determine whether the Tumblestick is properly balanced so that it will come back. Some sticks are erratic and turn in a different direction each time they are thrown —these should be discarded. If the stick performs in the same way each time, yet does not come right into your hands, a little experimentation should determine how it can be thrown so as to cause it to reach you. For example, if the Tumblestick, when thrown straight ahead of you as described, returns so far to the left of you that you cannot catch it, try facing straight ahead but throwing it at an angle a little toward your right side. Since it has a tendency to float to the left, this should bring it back to your hands. Similarly, if the stick tends to drift too far to your right, throw off in an oblique direction to your left.

Remember that no two boomerangs of any type can be thrown in just the same way. We must experiment with each one to find out what its peculiarities are, and then take pains to throw it in this way each time.

CHAPTER VI

Australian Boomerangs—How to Make Them

While boomerangs of the Australian type will not prove to be as interesting for general recreational use as the types described in the preceding chapters, yet a person who becomes interested in boomerangs as a hobby will certainly want to familiarize himself with them. There are several reasons why the Australian type is not so satisfying and is not recommended for general use: In the first place, these boomerangs are difficult to make—they call for heavy wood that is hard to work, and require a twist or skew which frequently exhausts the patience of the amateur. Secondly, the best of them are not as accurate in their return as the ordinary Cross-sticks and Pin-wheels which can be made in a few minutes. The primary argument against them, however, centers around the greater element of danger in their use: They are heavy and must be thrown with great force, consequently extreme caution must be taken to safeguard against injury to people or property. Their weight and shape make it unwise to attempt to catch them in any way except with the use of a net.

In favor of the Australian Boomerangs, it can be said that their use is excellent exercise. They must be thrown out-of-doors, and one must throw them hard. The fascination of attempting to make them return is such as to furnish incentive enough to keep one at it for a long time.

Certainly an adult or older boy who takes up the boomerang sport seriously will want to learn to make and throw these curved missiles perfected by the far-off Australian Bushmen. As stated in Chapter I there are two types of Australian Boomerangs, the *return type*, and the *non-return type*. Each of these will be described in turn. The methods of throwing Australian

Boomerangs are described in Chapter VIII, "How to Throw Boomerangs."

THE RETURN TYPE

The Australian Bushmen use acacia wood for the making of their boomerangs, but hickory or hard maple will answer the purpose very well. It is a waste of time to try to make boomerangs of this type out of light, soft wood.

The boomerang consists of two straight arms as illustrated in

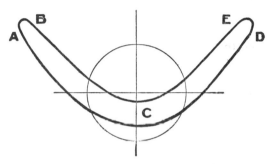

FIGURE 33. DIAGRAM OF AN AUSTRALIAN BOOMERANG

Figure 33. The circle indicates the center of gravity and the rotation of the missile around this center when the boomerang is in flight. It is possible to make boomerangs of the returning type with arms at angles of 70 to 120 degrees, although one seldom sees one with an angle of less than 90 degrees. Those made at angles of 110 to 120 degrees are the most common and the most convenient to use. Sometimes both arms are of the same length, and again one arm is an inch or two longer than the other.

One side, the bottom, remains flat whereas the top side is rounded off to a convex shape. The arms are bent slightly so that the ends (A B and D E in Figure 33) are raised above the plane of the boomerang at C as it lies on its flat side. In addition, each arm has a *skew* also; that is, the extreme end of each arm is twisted two or three degrees from a plane running through the center, so that B is below the plane and A above it, and D below it and E above it. This skew is an all-important factor and no curved boomerang will work without it when thrown from the vertical position.

These boomerangs may be made in lengths varying from eighteen inches from tip to tip up to three feet or longer. For our purpose it is recommended that an eighteen-inch or two-foot

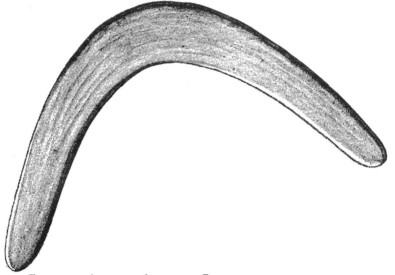

FIGURE 34. A TYPICAL AUSTRALIAN BOOMERANG OF THE RETURN TYPE

length be used. The primitive Australian Boomerang that was used for the model of Figure 34 measures twenty-one inches from tip to tip and weighs one pound. It is three inches wide at the bend, one-and-seven-eighths inches at one end, and one-and-one-half inches at the other end.

Although the proportions vary somewhat, it is a safe guide to say that the length of the wing or arm is six times its width at the widest point, and that the width of the wing is six times its thickness at the thickest point. Most primitive Australian boomerangs will not vary far from these proportions.

To make a boomerang out of one piece of wood is a most difficult task, in that it is next to impossible to bend the stick permanently into the proper angle. Once in a long time one may find in the woods a hardwood branch or root which has just

FIGURE 35. A BOOMERANG MADE BY PRESENT-DAY NATIVE AUSTRALIANS, SHOWING BURNT DECORATIONS

the right bend so that if it were stripped up, it would furnish sticks already possessing the angle necessary for a boomerang. If one were to set out to find one of these, however, he might hunt for months without success. The only other method would be to steam or soak a strip of wood and then bend it to the proper angle, keeping it in the bent position until it dries. This requires apparatus that the average person does not possess.

It is possible, however, to make an Australian-type boomerang out of two pieces of wood joined together at the bend. This is the method which we shall describe. If one is fortunate enough to secure a curved piece of wood, the wood can be worked and given the skew in just the same way that will be described for the one made of two pieces.

First make a pattern from a piece of paper twenty-one inches long by twelve inches wide. This pattern is shown in A, Figure 36. Drawn as indicated, the arms should be sixteen inches long, two-and-one-half inches wide near the angle, and one-and-one-half inches wide at the ends. Cut out the pattern, and draw two lines across it near the bend as shown in B, Figure 36.

Place the pattern on a piece of hickory, hard maple, or ash, five-sixteenths inch thick, and trace each arm of the boomerang as shown in E, Figure 36. Cut out each arm with a keyhole saw. The curved section of each arm (between the lines on the paper pattern) must now be cut into a half-lap joint as shown in C.

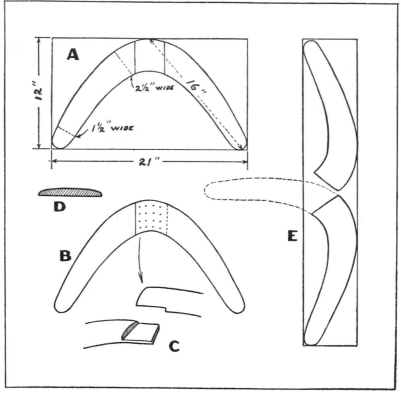

FIGURE 36. PLANS FOR MAKING AN AUSTRALIAN BOOMERANG

This can be done with a hacksaw or miter saw. Glue this joint and set in a press to dry. Then drive in some small nails to give additional support.

Now with drawknife, wood rasp, and jackknife, bevel off the edges on the top side of the boomerang and round off to a convex shape, as indicated in the diagram of the cross section in D, Figure 36. The stick remains at full width at the center, but is pulled down to a feather line at the edges.

The result now has all of the appearances of a boomerang yet it lacks the essential features that cause it to return. The two secrets here are a bend in each arm, and a skew near each end. Both the bend and the skew are so slight that the average person does not detect them in looking at the boomerang.

The Dihedral Angle.—The arms should be bent before they are skewed. The bend consists of a very slight turn upward, that is, toward the convex side. To accomplish this, hold the arm of the boomerang over a flame so that the flame hits it at a point two-thirds of the distance from the end to the curve. When the wood is hot place the heated spot against the edge of a table and press hard, thus bending the arm slightly. The bend thus caused will remain permanently. Bend the other arm in the same way.

After these bends have been properly made each arm will possess a slight dihedral angle. When it is placed on the bench with the flat side down, the ends of the arms will be slightly higher than the middle section of the boomerang. This angle is so slight, however, that a casual observance of the boomerang would not detect it.

The Skew.—It is now necessary to give each end of the boomerang a slight skew. Referring to Figure 33, which represents the convex side of the boomerang, the end must be so skewed that A is above the plane running through the center and B is below it. Similarly the other end must be so skewed that E is above the plane and D below it.

To accomplish this skew, heat the end of the boomerang over a gas burner for a distance of about four inches. When hot, grip the end with a pair of heavy pliers held in one hand, and grip the arm four inches from the end with another pair of pliers held in the other hand. Bend the arm two or three degrees in the desired direction and hold for a moment or so.

Now throw the boomerang and if it does not return properly

increase the skew a little. Continue to experiment by increasing and decreasing the skew until you get the results you want.

Very seldom does one see a description of how boomerangs are made which makes any mention of the bend in the arms or of the skew. Both are essential. It is absolutely impossible to make a boomerang of this type return or even swing to the left with an indication of returning unless the arms are given a skew in the direction indicated above. The secret of the curved boomerang of the Australian type is in the skew.

THE NON-RETURN TYPE

The non-return type of boomerang employed by the Australian Bushmen for hunting and warfare will travel for great distances with uncanny accuracy. As made by the Bushmen, these boomerangs vary from six inches in length to four feet. The big boomerang from Australia which was used for the model

FIGURE 37. A BIG HUNTING BOOMERANG OF THE NON-RETURN TYPE

of Figure 37 measures exactly four feet in length and weighs three-and-one-fourth pounds; it is two-and-one-fourth . inches wide at the middle section, this width remaining constant through the greater length of the boomerang; the tapering starts near the ends and comes down to one-and-three-fourths inches near each tip.

It will be noted that most boomerangs of this type are sickle-shaped and do not possess the sharp angle that is characteristic of the return type. This is not an essential feature, however, in that the models with a sharp curve will, if made correctly, perform in the same way.

Non-return boomerangs are made by exactly the same method

described for the return-type with one important exception: the skew at the end of each arm is in the reverse direction from that found in the return type. That is, B in Figure 33 is higher than the plane of the boomerang and A is lower; similarly D is above the plane and E below it. The skew is made by the method already described for the return boomerang.

When thrown in a position vertical to the ground this type of boomerang will travel for a long way without swerving to the right or left. It can be made to come back, however, if it is thrown from a position parallel to the ground as described on page 87, Chapter VIII. In this case it rises at a steady angle to a great height and then glides and volplanes down to the thrower.

CHAPTER VII

Miniature Boomerangs of Cardboard

Like the circus, the little boomerangs of cardboard are supposed to be stuff for the amusement of children, but curiously enough they find an equally robust following among grown-ups, which latter usually make them on the pretext of amusing the children but in fact are equally intrigued by them themselves. Certainly any one, old or young, who finds a hobby in boomerang making and throwing will put in happy hours with these little cardboard floats, while sitting around the house in the evening. Requiring but a few feet of space, they can be safely twirled in any room of the house.

A clever turn may be given to a boomerang-throwing act or demonstration by inserting one or two of these tiny boomerangs between the throws of the big and far-flying wooden boomerangs. The sudden change in size from the huge to the tiny is always a good trick of the stage.

A pair of scissors and a piece of cardboard will make boomerangs of every style described in these chapters, and the making is a task of but a moment or two. Any kind and weight of cardboard may be used, ranging from the thin pasteboard back of a writing tablet up to the very heavy board, the only requirement being that the pasteboard be solid and not of the corrugated type.

The heavier the cardboard the larger the boomerang may be. The dimensions given in this chapter are for pasteboard of the weight customarily used on the back of a writing tablet. If heavier cardboard is used, increase the dimensions accordingly.

CARDBOARD CROSS-STICK BOOMERANGS

On a piece of cardboard eight-and-one-half inches square, mark out the cross shown in Figure 38, making the wings one inch wide. Then cut it out with a pair of scissors. Give each wing a slight bend upward with the fingers at a point about two inches from the end. Do not bend it enough to break or crease

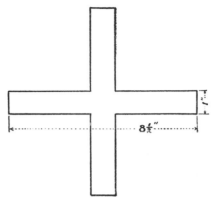

FIGURE 38. A CARDBOARD CROSS-STICK BOOMERANG

the cardboard, but rather just enough to give it a slight curve. This is all there is to it and the boomerang is now complete.

Now to throw it: Hold the boomerang between the thumb

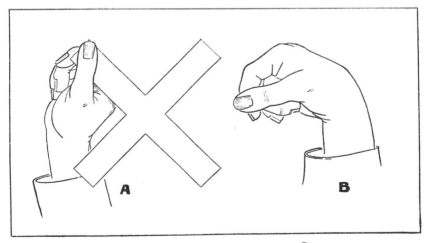

FIGURE 39. THROWING THE LITTLE CARDBOARD CROSS-STICK

and forefinger and turn the wrist back as in A, Figure 39. Throw it by snapping the wrist forward quickly and as you do so let the boomerang rotate forward between the thumb and forefinger so as to give it increased spinning motion.

The boomerang will sail forward a few feet, swing to the left and rise higher in the air, turn over into a horizontal plane, and then glide and volplane back into your hands. The performance is strikingly like that of a heavy cross-stick boomerang made of wood.

A space of twelve or fifteen feet square is ample room in which to throw these tiny boomerangs.

Larger boomerangs of this type may be made of heavier cardboard in just the same way.

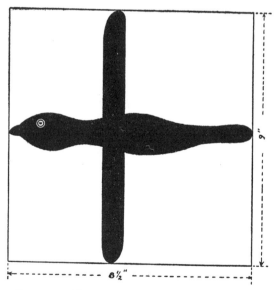

FIGURE 40. DIAGRAM FOR THE CARDBOARD BOOMABIRD

CARDBOARD BOOMABIRDS

Boomabirds, too, may be made from cardboard. Draw the bird on a piece of cardboard measuring eight-and-one-half inches wide by nine inches long, as illustrated in Figure 40. The body of the bird is one-and-three-eighths-inches wide at the widest point, and eight-and-one-half inches long. The wings are about an inch wide at the widest point and measure nine inches from tip to tip. Cut out the bird with scissors and bend the end of each wing, also the head and the tail, upward slightly as was done in making the cardboard Cross-stick described above.

The Boomabird is thrown in just the same way as is the cardboard Cross-stick. Its flight follows a course that resembles quite typically that of a large wooden Boomabird.

It is also possible to propel the Boomabird by snapping it from a book, following the method described for the cardboard Australian boomerang in the following section. Its flight in this case is very different than if thrown with the hand.

CURVED CARDBOARD BOOMERANGS

A cardboard boomerang resembling in outline the curved boomerangs used by the Bushman of Australia is illustrated in Figure 41. When made of light cardboard, the arms should be an inch wide, one of them five inches long and the other about a half inch shorter. No bend or curve in the wings is necessary in this boomerang. Merely cut it out and it is ready to throw.

Lay the boomerang on a book allowing one end to project over the

FIGURE 41. "THROWING" A CURVED BOOMERANG MADE OF CARDBOARD

edge about an inch, as shown in Figure 41. Hold the book containing the boomerang in the left hand and a pencil in the right, as illustrated. Give the boomerang a sharp rap with the pencil, thus sending it forward and upward. Or, if you choose, discard the pencil and snap it with the finger.

The boomerang will sail forward and upward eight or ten feet, turn and glide directly back in practically the same plane in which it rose.

It will be noted that this boomerang is not thrown from the vertical position as is the pasteboard Cross-stick Boomerang, nor can it be thrown in this way. Consequently the course of its flight is different. When snapped from the horizontal position, its flight resembles that of a heavy Australian-style boomerang of the non-return type when such a wooden boomerang is thrown from a horizontal position. That is, it sails forward and upward, the boomerang remaining in a horizontal position, and then volplanes back down.

CARDBOARD TUMBLESTICKS

Excellent Tumblesticks may be made from lightweight cardboard. Using a length of eight-and-one-half inches, the Tumblesticks may be made in the following widths: one-and-seven-eighths inches, one-and-three-eighths inches, one inch, and one-half inch. The wider they are the slower they rotate when thrown! When heavier cardboard is used the length and width may be increased accordingly.

No description of how to throw cardboard Tumblesticks is necessary here in that they are thrown in precisely the same way as the large wooden Tumblesticks described on page 62.

If one fails in his efforts to make a wooden Tumblestick return, and thus reaches the decision that a straight missile cannot be thrown so that it will come back, all that is necessary to convince him is to have him try one of the cardboard Tumblesticks. If properly thrown as described in Chapter V, they are practically "fool proof."

CHAPTER VIII

How to Throw Boomerangs

All the sport of boomerang making is but preliminary to the joy of throwing. And any one who has never made a boomerang and thrown it for the first time, can scarcely realize the thrill that results as it comes floating back, circles over his head, and settles in his hands. It is worth many times over the labor required for the making.

Most people have the idea that skill in throwing boomerangs can be developed only by long and arduous practice, yet such is far removed from the fact. The skill rests in the making more than in the throwing. Given a perfect boomerang, any novice of average athletic ability will find himself catching it regularly after a half hour of practice, provided the instructions in this chapter are carefully followed.

There are of course many "tricks of the trade" in boomerang throwing which are acquired only from experience, but such is true in any sport. Each boomerang offers a problem all its own, and varying conditions of space and weather contribute added complications. The veteran sizes up these factors at a glance, whereas the novice must learn by trial and error. But after all is said and done, given a good boomerang and proper air conditions, success comes more quickly than in many sports that might be mentioned. However, it is absolutely necessary that careful attention be given to the fundamental techniques described in the following pages—the veteran performs these without giving conscious attention to them whereas the beginner must study each movement.

For some unexplainable reason, boomerangs will return accurately when thrown so as to curve from right to left but cannot be caused to return satisfactorily or consistently from left to right. It would seem that a good Pin-wheel, for example, which works perfectly when thrown so as to curve to the left, would, if turned around, circle to the right with equal accuracy. Such, however, is not the case, and no one has been able to offer a satisfactory explanation. A left-handed person would naturally reverse the position of the boomerang and throw it with the left hand, and the normal flight of such a throw would be from left to right. Within my experience, however, I have never found a person who was consistently successful with such a throw. The result is that left-handed people either throw with the right hand, or hold the boomerang in the left hand in such a way as to cause it to swing to the left. Occasionally a boomerang is found that can be made to fly in both directions, but after experimenting with countless boomerangs of all types over a long period, the conclusion must be reached that *boomerangs perform accurately and consistently only when thrown so as to curve from right to left.* Mysterious, indeed, but it is a fact nevertheless!

All boomerangs except the Tumblesticks are thrown in practically the same way. Consequently, the instructions in this chapter will apply in a general way to all types—Cross-sticks, Pin-wheels, Boomabirds, and curved Australian Boomerangs. Where there are exceptions, attention will be called to the fact. The instructions for throwing Tumblesticks are given in Chapter V, "Tumblesticks—How to Make and Throw Them."

Before any thought can be given to the methods of throwing, we must first think about any elements of danger that may be involved. No one would want to be responsible for hurting a bystander or injuring property, not to mention breaking his boomerang which probably would be the result if it collided with an object.

AVOIDING ACCIDENTS

First off, let the fact be clear that this book does not recommend the general use of curved boomerangs of the Australian type. The Cross-sticks, Pin-wheels, and Boomabirds are safe and harmless when used with a little ordinary common sense such as is exercised in archery, swimming, horseback riding, shot putting, and scores of other sports. The heavy Australian Boomerangs are dangerous except when used by mature and careful people. Their use should be confined to large, open areas, and to times when none but the throwers are present. Except under very careful supervision, they have no place on playgrounds, in camps or similar areas. *Substitute for them the light and safe Crosssticks and Boomabirds—they are more efficient and more enjoyable anyway.*

Assuming that the boomerangs are of the type recommended in these chapters, there is no more danger in their use than in most other sports provided the same degree of precaution is exercised. There is some danger in any game in which a ball is thrown, but this does not prohibit ball games—neither should it rule out boomerang throwing. With common sense in the driving seat, boomerang throwing may safely take place either in a gymnasium or outdoors.

In the interest of safety, common sense dictates that the following regulations should always be carefully observed:

1. Always have the spectators and bystanders gather in a group to your right. The boomerangs curve from right to left, and consequently offer no danger to those who are gathered to the thrower's right. If a boomerang does reach them it will be on its return trip and its force will be spent—it will settle down gently and harmlessly.

2. *Never throw when people are scattered about the area,* even though they all know the throw is coming and are watching. Since a boomerang curves, the spectators cannot always judge which way it will turn. Even though these light boomerangs probably would not do serious damage, *safety first* is the rule always. An expert performer can manipulate boomerangs around people, even to the point of throwing them from the stage of a theater, but such stunts are not for novices.

3. *Never throw without first yelling a warning* so that every one is informed and can watch.

4. *Never throw a boomerang in a strong wind.* The wind will carry it far from where you intend it to go.

5. *Tighten the bolt securely before throwing*—if it loosens, the boomerang may fall unexpectedly.

6. *Never throw a boomerang near a building*—it may find a window.

THROWING

The would-be boomerang thrower immediately comes face to face with a most serious problem: the wind blows his boomerangs all over the lot. And if there is much breeze blowing, there is no way to prevent such a happening. Some of the heavy curved boomerangs of the Australian type are at their best in a breeze since they are not accurate enough to return without the help of the wind, but not so with the Cross-sticks, Pin-wheels, and Boomabirds—these are made to return perfectly when the air is calm, and so cannot be expected to stay true to their course when buffeted by the breeze. Even an expert performer would appear to a bad advantage if forced to work in the wind.

For perfect performance and complete satisfaction, a large gymnasium is the boomerang thrower's dream. All but the large and heavy boomerangs can be thrown in the average-sized school gymnasium. In the theater, the circus tent, and the gymnasium there is no wind to defeat us. Given a calm day when the air is still or the breeze very slight, there is great joy to be had in outdoor boomerang throwing, particularly in using the Boomabirds. Usually there is less movement of the air in the evening than at other times of the day. When a new boomerang is being tried out, it is better to test it in a gymnasium or when there is perfect calm outdoors; otherwise one cannot tell whether it is made just right or not.

This leads us to the first two important rules of throwing boomerangs outdoors: First, pick a calm day when there is little or no breeze; and second, always throw directly into the breeze— handled in this way, the devastating effect of a slight breeze can be quite effectually overcome. The boomerang of course moves with greater force on its outward journey than on its return. When thrown against the wind, the wind assists rather than blocks it in making its turn. An absolutely calm day very seldom occurs. The leaves may be motionless and the lake like a mirror, yet there are air motions that a delicate boomerang will instantly register. The direction of the breeze may be located by wetting one's finger and holding it up in all directions. Better still, throw a light boomerang of perfect action once in every direction. This will settle the matter beyond controversy—it will come back more perfectly when thrown in one certain direction and that direction should be used for all throwing at that time.

THE STANDARD OR VERTICAL THROW

Throwing the Cross-sticks and Boomabirds.—Hold the boomerang in the right hand, gripped rather loosely but securely as near the end of one wing as possible. Place it in the straight up-and-down position shown in B, Figure 42, directly perpendicular to the ground. The beveled or convex side must be toward you. If you are left-handed, either try to throw the boomerang from your right hand, or hold it in the left hand with the beveled side *away* from you.

From the position shown in A, Figure 43, draw the arm back to the position in B, being careful not to tilt the boomerang sidewise. Throw it straight forward as in C, and just as you release it, turn the wrist and forearm sharply downward as in D, the latter movement being made in order to give the boomerang as much spin as possible. A great deal of muscle is not necessary, but it is usually essential that you give it an emphatic spin by turning your wrist downward in releasing. The entire throwing movement is more one of the forearm than of the entire arm, especially in handling boomerangs of ordinary size.

If the boomerang fails to return as it should, it may be that you are unconsciously turning it or "slicing" it at an angle in releasing it. This is a common fault among beginners. People who play tennis seem to be particularly troubled in this respect at the start—they are often inclined to give their forearm a slight turn as in cutting the ball while serving in tennis. This of course turns the boomerang and the throw goes awry. Check to see that you are throwing straight forward and releasing the boomerang without tilting or twisting it.

If, after a thorough trial, the boomerang does not return, try holding it at one of the angles shown in A and C, Figure 42.

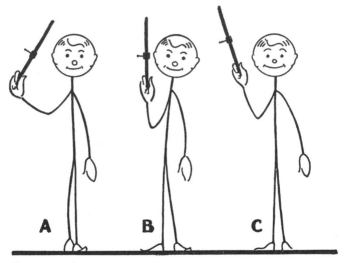

FIGURE 42. ANGLES AT WHICH BOOMERANGS ARE HELD IN THROWING

Which of these angles to use is determined by how the boomerang acts when thrown from the vertical position. If the boomerang cuts in too soon each time and falls in front of you, it must be thrown from the position in C—tilt it slightly *away* from you when you throw it and it should return directly to you. On the other hand, if the boomerang does not cut in soon enough and is of the type that persists in swinging around behind you and falling to the ground, it should be thrown from the angle in A—tilt it slightly *toward* you.

Remember that no two boomerangs can be thrown in precisely the same way. We must become familiar with each boomerang and determine by experience just how it must be thrown for best results. It is no reflection on the worth of a boomerang if it cannot be thrown from the up-and-down position shown in B, Figure 42, but requires one of the inclined positions in A and C. Many of the finest of boomerangs require one of these slanting throws. Once you discover the proper angle, make a note on the back side of the boomerang so that you can tell at a glance what to do with it.

Now the type-perfect Cross-stick or Boomabird must be

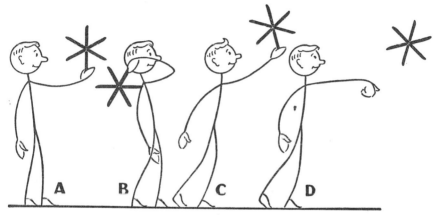

FIGURE 43. THROWING THE PIN-WHEEL, ILLUSTRATING THE VERTICAL THROW USED FOR MOST TYPES OF BOOMERANGS

thrown straight out at the level of the head and parallel to the ground. It will rise higher of its own accord as it starts to turn and will then settle down within easy reach of the hands on its return. If thrown higher into the air, it will sail over the head and out of reach. However, individual boomerangs differ as much here as in any other respect. Some must be thrown at an angle of at least forty-five degrees into the air. Experimentation is the only rule in discovering the proper angle, and once it is determined, it should be noted with pencil on the back of the boomerang.

Similarly, some boomerangs must be thrown with great force,

some with average speed, and others very gently.   And once this
item is ascertained, a note to that effect should be made on the
boomerang.

There are thus three items that should be recorded with pencil
on the back side of the boomerang: (1) the angle at which it
should be held, whether A, B, or C, as shown in Figure 42; (2)
the angle at which it should be thrown into the air, whether high,
medium or low; (3) the speed with which it should be thrown.
The notations are not only a great convenience always, but are
indispensable when one is throwing a number of boomerangs, one
after the other, as in an exhibition of boomerang throwing.   A
glance at the back and we know definitely how to handle each
boomerang and what to expect of it.   The spectators of course
are not aware that any notations are being used.   Since typical
performance would require the straight up-and-down position
(B, Figure 42), average speed, and a throw straight out and
parallel to the ground, these can be taken as standard, and notations made only on the boomerangs that depart from this pattern.
If no notations are to be seen, we assume that the boomerang is
held and thrown in the usual way.   And again, if only one of
the three items is noted, the assumption is that the boomerang
is handled typically in the other respects.   Thus, if the penciled
notation reads "C, *high, easy*" we know that it must be tilted
slightly outward, and thrown gently at a high angle into the air.
Whereas if the notation merely said "*A*," we assume that if it is
held so that it is tilted slightly inward, it will come back if
thrown straight out and with average speed.

Once in a long time a boomerang that is usually dependable
will fly forward a few feet, turn suddenly to the right, and nose-
dive to the ground.   This is called "back firing" and is a bad
fault against which definite precautions must be taken.   It is
due to the fact that at the moment of release the boomerang was
held so that the front wing was pointed a little to the right rather
than straight ahead, and the force of the throw caused the air
to catch it and turn it to the right.   Boomerangs two feet in size
or smaller can usually be grasped between the thumb and fore-

finger at a point an inch or two above the end of the wing. Held in this way, the pressure of the thumb against the bevel of the wing turns the front wing a little further to the left and prevents any possibility of "back firing." With experience, a person develops the knack of throwing the boomerang in such a way that "back firing" seldom if ever occurs. However, if a boomerang "back fires" often, *discard it*. It could not have a worse fault. Like some horses, some boomerangs will always remain obstreperous and persist in bucking when you least expect them to. Such are good only for kindling wood.

It is the glory of a good boomerang that you can depend upon it to travel in precisely the same way every time it is sent forth.

Throwing the Curved Australian Boomerang.—The curved Australian Boomerangs of the *return* type are thrown just as described above except that they must be hurled with great force as compared to the other types, and are given the spin or revolving motion in a different way. Just as the boomerang is released, the hand is given a sharp upward jerk. This sharp jerk is important because without it the boomerang will spin lazily at the start and will not have enough revolving motion to make the return journey. Most Australian Boomerangs will perform correctly if thrown from the straight up-and-down position illustrated in B, Figure 42. It is seldom necessary to throw from the positions shown in A and C.

The average boomerang of the return type will go forward thirty to fifty yards before swinging to the left, then rise high in the air, and volplane back.

The *non-return* type is thrown in just the same way if the object is to hurl it accurately at a target some distance away. It is possible to throw these non-return boomerangs so that they will come back, but this method is described in the following section on the horizontal throw.

THE FLAT OR HORIZONTAL THROW

Throwing the Cross-sticks and Boomabirds.—Hold the boomerang in the right hand parallel to the ground, that is, at an angle of forty-five degrees in respect to the ground, with the beveled or curved side up, as shown in Figure 44. Throw it forward and upward, giving it as much spin as possible with the wrist in releasing it. It will sail high up into the air, hesitate, and then glide and volplane back in practically the same plane as in the ascent. This is not as spectacular a throw as the vertical but it is interesting as a variation. Not all boomerangs will work when thrown in this way. Occasionally a balanced Cross-stick is

FIGURE 44. THE FLAT OR HORIZONTAL THROW

found that can be thrown from either the vertical or horizontal position. Usually, however, a Cross-stick designed to be thrown from the horizontal position should be made with shorter and wider wings, and the wings should have only a very slight upward bend or none at all. An ideal size is made of two sticks eighteen inches long, one-and-one-fourth inches wide and one-eighth inch thick.

Some boomerangs must be thrown in a way that is a cross between the vertical and horizontal methods. They must be tilted at a wide angle away from you, much wider than that illustrated in C, Figure 42, yet not so far as to be in the horizontal position shown in Figure 44. The Two-wing Cross-sticks (Figures 15, 16, and 17) and the Fans (Figure 24) are of this type. In flight, these boomerangs follow the typical course of the boomerangs thrown from the horizontal position—they sail out and up at an angle of about forty-five degrees, then volplane back in practically the same plane.

Throwing the Curved Australian Boomerangs of the Non-Return Type.—The curved Australian Boomerangs of the *non-return* type may be caused to return by throwing them from the horizontal position. When thrown from the *vertical* position they go straight forward with great accuracy, wavering not at all to the right or left, but when thrown from the *flat* or *horizontal* position, they go high up in the air and then glide back to the thrower in essentially the same plane. These boomerangs have a reverse or backward skew in the arms, as described in Chapter VI, which accounts for the difference in action.

Frequently one finds in the stores commercial boomerangs supposedly of the *return* type which have all the appearances of good Australian Boomerangs but which fail to return when thrown in the normal way from the vertical position. These boomerangs usually are at fault in that they do not have a skew in the arms. As a rule these can be thrown somewhat successfully from the flat or horizontal position.

CHARACTERISTIC FLIGHTS OF BOOMERANGS

No two boomerangs perform in exactly the same way. Each has a temperament of its own and certain habits of flight that give it a personality differing from all others. This is one of the chief reasons why people become so thoroughly absorbed in boomerangs as a hobby. They keep on making more and more, knowing that each one will differ slightly from all others in its ways of behaving, and that these ways cannot be anticipated accurately until the boomerang has been thrown and its habits studied.

Figure 45 shows typical courses followed by boomerangs in their flight; there are of course endless variations of each of these.

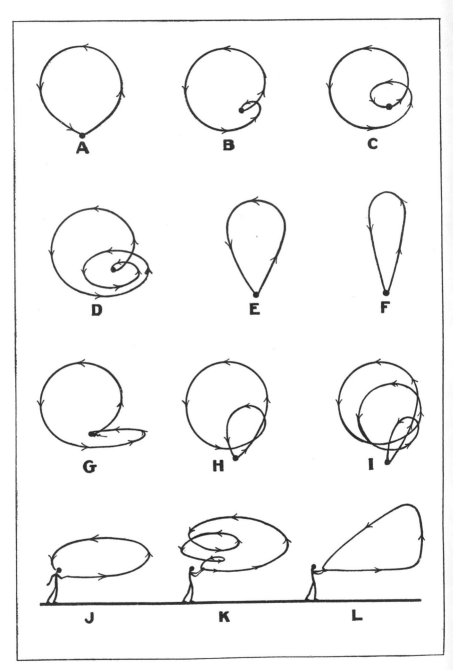

FIGURE 45. CHARACTERISTIC FLIGHTS OF BOOMERANGS

Some boomerangs are inclined toward circular flights as shown in Figures A to D, while others tend toward an oblong or somewhat of an egg-shaped flight as illustrated in E; still others go a long distance, turn around rather abruptly, and come back in almost a straight line, as diagrammed in F.

Light-weight Cross-sticks follow a circular course as a rule. Some cut a circle in front of the thrower and come in to his left side as shown in A. Others will swing around and come up behind him, or come in from his right side, as shown in B. Still others will make a circle and a half before reaching him and come in from his left side, as in C.

Light-weight Pin-wheels act much like the Cross-sticks. A, B and C illustrate their typical courses. Those with short wings are inclined to follow the patterns in A and B, whereas those with the long wings often stay in the air longer and follow the course shown in C.

A heavy Pin-wheel with short wings, similar to that illustrated in Figure 21 on page 43, frequently follows the delightful course shown in D, rising high in the air and circling three times before reaching the thrower.

Very heavy Cross-sticks and Pin-wheels with long wings tend toward the more oblong courses shown in E and F.

The Boomabirds with their wide bodies rotate slowly and stay in the air a long time, their flights forming the patterns shown in G, H, and I. G shows a very characteristic flight of a Boomabird when given a gentle throw. Forceful throws will often send them through the courses of H and I.

The curved Australian boomerangs go straight forward for a long distance, swing to the left and cut back, following a course similar to that shown in E.

Diagrams A to I in Figure 45 show how the flights of boomerangs would appear if one could look down on them from above. Boomerangs also perform differently, one from another, in the elevation from the ground that they reach in the course of the flight.

The light Cross-sticks and Pin-wheels go forward vertically

for some distance at the level at which they were thrown, then circle to the left, turn over on the flat side and rise slightly in the air, remaining at the higher level until they near the thrower. Viewed from a point to the right of the thrower, the flight appears as shown in J, Figure 45. The Boomabirds and the heavier Pin-wheels go forward in much the same way, but as they turn to the left they often rise high in the air, turn over on the flat side, circling and double circling, and finally settle down to the thrower —Figure K illustrates such a course as it would appear viewed from the right of the thrower.

The Australian boomerangs, although varying in detail, all follow a rather characteristic, parabolic course. They go forward a long way in almost a straight line, turn rather abruptly to the left and at the same time rise high in the air; at the highest point they turn over so that they are parallel to the ground, flat side down, and then glide and volplane down to the thrower. Figure L indicates this course as it would appear from the right side of the thrower.

CATCHING BOOMERANGS

No matter how accurately a boomerang returns, if the thrower does not catch it, the impression created is one of failure. To "muff" it is worse in its effect than if the thrower had to walk a few steps to reach it—the spectators may not notice the steps in that their eyes are probably on the boomerang, but they cannot fail to see the miss in catching.

The method used in catching depends upon whether the boomerang is held together with a long or short bolt. Given a short bolt, it is caught by the wings, whereas with a long bolt, it is caught by the bolt. There is a third method sometimes used for curved Australian boomerangs—that of catching them in a net. Nets are useless for other types of boomerangs. A trick method of catching boomerangs on the head is described later in this chapter under the section on "Trick Stunts."

CATCHING BY THE WINGS

Do not attempt to catch the boomerang with one hand. To do so presents two hazards: the odds are in favor of missing it in that the spinning wings are hard to judge and to secure, and

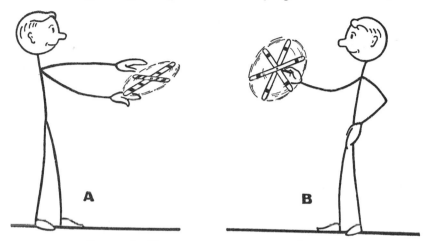

FIGURE 46. TWO WAYS OF CATCHING THE CROSS-STICKS

there is danger of bruising the fingers if they are thrust into the circling wings.

To catch it, *hold one hand above it and the other below it, and clap the hands together,* as shown in A, Figure 46. It is pretty hard to miss by this method if the boomerang is within easy reach, and the danger of injury is so negligible as to be practically non-existent. This method is safe and sure.

CATCHING BY THE BOLT

The presence of a four-inch bolt (see page 20) simplifies greatly the process of catching the boomerang, and at the same time adds much to the picture since the boomerang continues to spin after it is caught. The average spectator will not detect the bolt and consequently wonders how the boomerang is being held so that it continues to spin. True, the bolt is small and is often hard to see, but nevertheless it offers less grief and insures more consistent success than otherwise, and withal is more spectacular and picturesque. Happily, most Pin-wheels and Cross-sticks, as they approach you on their return, turn the flat or back side toward you, and consequently present the bolt in a position that is very easy to grasp.

As the boomerang comes in, keep your eye glued on the bolt but do not reach out with your hand until it is in position to catch. Merely hold your hand up and in readiness. When the boomerang is where you want it, shove your hand out quickly and grab the bolt firmly and emphatically. To reach out with your hand and follow the motion of the bolt with it often results in the boomerang getting away. When you are sure you know just where the bolt is, snatch it suddenly. In doing so, however, be careful not to stop the flight too abruptly; rather, let your hand move along with the boomerang a foot or so and then swing it around in front of you with a graceful curve and at the same rate of speed at which the boomerang was moving. The Pin-wheel in B, Figure 46, has just been captured by the bolt.

CATCHING WITH A NET

There is no way to catch an Australian Boomerang of the curved type in the hands with anything approaching safety. They are as dangerous to the thrower as to the spectators. Better let them alone entirely. However, if they are used, do not attempt to catch them in any way except with a net. Secure a net frame of the size of a large butterfly net, equipped with a three-foot handle. Attach a net made of heavy mesh, such as is used in tennis nets. If the Australian Boomerang comes close enough to you, snatch it in the net. Do not bother with nets in catching other types of boomerangs.

TRICK STUNTS

Here are some interesting trick throws that are excellent for use in boomerang-throwing exhibitions. These will require long practice but can be easily mastered by those who are familiar with the goings and comings of boomerangs and the techniques of throwing and catching.

THROWING TWO BOOMERANGS IN SUCCESSION

Pick out two Pin-wheels that are mates in size and in action, each having a long bolt with which to catch it. Hold one boomerang in each hand. Throw the boomerang that is in the right hand, immediately transfer the other to the right hand, and throw it as quickly as possible. The first boomerang will return a few seconds before the other—catch it by the bolt in the right hand and then catch the other in the left. Allow both boomerangs to continue to spin in your hands after catching.

This is not a difficult feat to learn. Be sure to throw the second boomerang just as quickly as possible after the first in order to give yourself time to get your bearings and get set for the catch. When both boomerangs are in the air, do not glance at the second until the first has been caught—if you do you will probably catch neither of them.

THROWING TWO BOOMERANGS AT ONCE

Select two small Cross-sticks that are perfect mates in size and action. Each should be held together with wire, rather than a bolt, as described on page 18. Lay one boomerang on top of the other and grasp both in the right hand. Throw the two just as if only one was being hurled. Shortly after they leave the hand they will separate and follow separate courses, but both will return and within a second or two of each other. If they are close enough together when they return (which isn't apt to happen), catch both at once by slapping them together with the hands. Otherwise, catch the first to return by clapping the hands on it, and, letting it lie on the lower hand, catch the second in the same way on top of it.

Sticks eighteen inches long, one-and-one-fourth inches wide, and one-eighth inch thick will make boomerangs that are ideal for this stunt.

FOUR IN THE AIR AT ONCE

This is the combination of the two preceding stunts. Four small Cross-stick boomerangs are needed of the size described in the foregoing paragraph for throwing two boomerangs at once. Arrange the boomerangs in pairs, one on top of the other, and hold one pair in the right hand and the other pair in the left. Throw the pair in the right hand, immediately transfer the pair in the left hand to the right, and throw them in the same way. Catch the boomerangs as they return by slapping the hands together on them. In catching, hold the left hand above the right, and let the boomerangs remain in the right hand after they are caught, catching each succeeding one on top of the others. To perform this one, you must be alert and do some fast grabbing.

FIGURE 47. CATCHING THEM ON YOUR HEAD

CATCHING A BOOMERANG ON THE HEAD

Secure a tin-can cover that is about five inches in diameter. Attach it to a piece of elastic that is just long enough so that when the cover is placed on the top of the head, the band will slip snugly under the chin. A rivet will hold the cover and elastic together.

Select a slow-floating Pin-wheel that returns at a height a little above the level of the head and settles gently down. It should be held together with a long bolt.

Place the cover on the top of the head and throw the boomerang in the usual way. As the Pin-wheel returns, place the head under it and catch the bolt in the can cover, as shown in Figure 47. It will continue to spin on top of the head with a most picturesque effect. The stunt requires a boomerang that habitually returns at just the right level. At best, it will require practice.

Fred Stone, the famous and versatile comedian, performed this stunt spectacularly with a heavy Pin-wheel thrown out over the audience in the theatre.

PLAYING CATCH WITH A BOOMERANG

In this interesting feat, two people stand a few feet apart. One throws a boomerang so that it circles and is caught by the other on its return trip. This stunt can easily be done on the stage—the performers stand on opposite sides of the stage, as far front as possible, and the one at the right throws the boomerang out over the heads of the audience so that it is caught by his partner on the left of the stage.

Much experimentation will be necessary to find the boomerang that will do this stunt. In trying out new boomerangs, we sometimes find one that for some reason will not return all the way back but drops far over to the left and behind the thrower. Such a boomerang should be excellent for this stunt. Then, too, there are the boomerangs that are perfect in action, but which circle all the way around the thrower before returning, passing him far to the left on the first backward flight. These also are usable.

Experimentation will also be necessary to determine just how far apart the two partners in the stunt should stand for any particular boomerang. This distance should be at least fifteen feet if the stunt is to be at all impressive.

BOOMERANG THROWING CONTESTS

Playgrounds, camps, clubs, and schools that feature boomerang making and throwing in their programs will find that a boomerang-throwing meet will be enthusiastically received and will serve as an excellent climax to the period devoted to the making and practicing. The following events will prove suggestive for such a competition.

In all these events, a participant should not be permitted to use boomerangs that he did not make himself.

The events should take place in a large gymnasium, or if outdoors, on a very calm day. Remember that there is usually less breeze in the evening than during the day.

ACCURACY THROWING CONTEST

Mark a circle about one foot in diameter on the floor or ground. With one foot in this circle, the contestant throws his boomerang and catches it when it returns. He is not permitted to remove his foot from the circle, although he may turn around at will in following the course of the flight.

Each contestant is given ten throws in succession and is awarded one point for each catch. The one making the highest score wins. In case of a tie, the winners throw again.

Any type of boomerang may be used and a contestant may change boomerangs during the contest if he chooses.

CONTEST IN THROWING TWO BOOMERANGS AT ONCE

One boomerang is placed on top of another and the two are thrown as one (see page 93). Each contestant is given ten attempts and scores one point each time he catches both boomerangs. To catch only one counts nothing. The conditions are the same as in the Accuracy Throwing Contest.

CONTEST IN THROWING TWO BOOMERANGS IN SUCCESSION

The contestant throws two boomerangs, one immediately after the other, and then attempts to catch them both, as described on page 93. Each is allowed ten throws and one point is scored each time both boomerangs are caught. In other respects, the conditions described for the Accuracy Throwing Contest apply.

LONGEST FLYING BOOMERANG

This contest determines the boomerang that can stay in the air the longest time. Some boomerangs float for a long time, circling and double circling before returning. These are the ideal type for this contest.

The event is best conducted out-of-doors. A stopwatch is used to record the time from the second the boomerang leaves the thrower's hand until he catches it. Since this is a contest between boomerangs, a contestant may enter several boomerangs, if he chooses. Each contestant is given five throws with each boomerang, and is credited with his best time. The thrower must keep one foot in a circle twelve inches in diameter, as in the Accuracy Throwing Contest above, and if he fails to catch the boomerang legitimately, the time is not recorded although it counts as a trial.

FARTHEST TRAVELING BOOMERANG

This contest is to determine the boomerang that can travel the greatest distance from the thrower and then return to him so that he can catch it. Of course it is not possible to do more than to estimate the distance approximately. Establish a throwing spot one foot wide on which the contestant must keep one foot. The judges are scattered parallel to the line of travel. When the contestant throws, the judges mark the farthest spot over which the boomerang passed in its flight. Each is given five throws and is credited with his greatest distance. No throw is measured unless the thrower catches the boomerang.

"PLAYING CATCH" CONTEST

Two partners "play catch" with a boomerang, throwing it from one to the other, as described on page 95. The partners stand as far apart as they choose, provided there is at least fifteen feet between them. One throws the boomerang so that it reaches his partner on its return trip, who catches it. Ten throws are made and one point is credited each time a throw is caught. The pair with the highest score wins.

BOOMERANG WITH MOST UNIQUE ACTION

Some boomerangs perform peculiar stunts in the air and follow unusual courses, stunts that other boomerangs will not do. This event is won by the contestant whose boomerang, in the opinion of the judges, performs the most unique stunts in the air, or completes the most circles before returning. Each participant continues to throw until the judges are satisfied as to the characteristic action of the boomerang.

BEST STUNT WITH A BOOMERANG

This contest is won by the contestant who, in the opinion of the judges, performs the most unusual and unique stunt in throwing and catching a boomerang. Such stunts may be used as catching the boomerang on the head, throwing and catching while sitting down or lying in a prostrate position, having four boomerangs in the air at once and catching them all, and so forth.

MOST UNIQUE TYPE OF BOOMERANG

Boomerangs may be made in all sorts of unusual and bizarre shapes, based upon the standard types described in the preceding chapters. This contest is won by the contestant who, in the opinion of the judges, displays the most unique type of boomerang. No boomerang is considered unless the thrower can catch it without stepping, at least five times out of ten trials.

BIGGEST BOOMERANG

The biggest boomerang that can be caught by the thrower without stepping wins this contest.

SMALLEST BOOMERANG

This contest is won by the contestant who displays the smallest boomerang made of wood, provided he demonstrates to the satisfaction of the judges that it will come back to him when thrown. The boomerang must be thrown and not snapped from a book or other object (see page 75).

Boomerangs, 1-98
  Australian-style, 64-71
  Avoiding accidents with, 78-80
  Bending wings, 16-17
  Bird-shaped, 48-57
  Bolts for, 21
  Boomabird, 48-57
  Boomaplane, 55
  Cardboard, 72-76
  Catching by bolt, 91-92
  Catching by net, 92
  Catching by wings, 91
  Catching on head, 94
  Contests, 95-98
  Cross, The, 29
  Cross-stick, 11-36
  Curved wing, 23
  Decorating, 33-36, 56
  Definition of, 6
  Dihedral angle, 68
  Dimensions for Boomabirds, 53
  Dimensions for Cross-sticks, 24-26
  Dimensions for Pin-wheels, 41
  Fan, The, 46
  Fish-shaped, 55
  Flared-end, 26-27, 41-44
  Flat throw, 85-87
  Flights of, 87-90
  Flying fish, 55

  Four-wing, 14-29
  Heavy Pin-wheel, 40
  History of, 8-10
  Horizontal Throw, 85-87
  Ice-tongs, The, 32
  Illuminating, 34
  Jumbo Cross-stick, 24
  Jumbo Pin-wheel, 39
  Miniature, 72-76
  Non-return type, 69
  Painting, 32
  Pin-wheel, 36-47
  Razor, The, 33
  Return type, 65-70
  Skew, 68-71
  Square, The, 31
  Streamers for, 34
  Tapered ends, 27-29
  Three-wing, 44-47
  Throwing, 59-61, 77-98
  Thunderbird, 54
  Tools for, 13
  Trick stunts, 93-98
  Tumblesticks, 58-63
  Two secrets of, 22
  Two-wing, 29-33
  Types of, 7
  Vertical Throw, 81-85
  Wood for, 12

A CATALOG OF SELECTED DOVER
BOOKS IN ALL FIELDS OF INTEREST

DRAWINGS OF REMBRANDT, edited by Seymour Slive. Updated Lippmann, Hofstede de Groot edition, with definitive scholarly apparatus. All portraits, biblical sketches, landscapes, nudes. Oriental figures, classical studies, together with selection of work by followers. 550 illustrations. Total of 630pp. 9⅛ × 12¼.
21485-0, 21486-9 Pa., Two-vol. set $25.00

GHOST AND HORROR STORIES OF AMBROSE BIERCE, Ambrose Bierce. 24 tales vividly imagined, strangely prophetic, and decades ahead of their time in technical skill: "The Damned Thing," "An Inhabitant of Carcosa," "The Eyes of the Panther," "Moxon's Master," and 20 more. 199pp. 5⅜ × 8½. 20767-6 Pa. $3.95

ETHICAL WRITINGS OF MAIMONIDES, Maimonides. Most significant ethical works of great medieval sage, newly translated for utmost precision, readability. Laws Concerning Character Traits, Eight Chapters, more. 192pp. 5⅜ × 8½.
24522-5 Pa. $4.50

THE EXPLORATION OF THE COLORADO RIVER AND ITS CANYONS, J. W. Powell. Full text of Powell's 1,000-mile expedition down the fabled Colorado in 1869. Superb account of terrain, geology, vegetation, Indians, famine, mutiny, treacherous rapids, mighty canyons, during exploration of last unknown part of continental U.S. 400pp. 5⅜ × 8½. 20094-9 Pa. $6.95

HISTORY OF PHILOSOPHY, Julián Marías. Clearest one-volume history on the market. Every major philosopher and dozens of others, to Existentialism and later. 505pp. 5⅜ × 8½. 21739-6 Pa. $8.50

ALL ABOUT LIGHTNING, Martin A. Uman. Highly readable non-technical survey of nature and causes of lightning, thunderstorms, ball lightning, St. Elmo's Fire, much more. Illustrated. 192pp. 5⅜ × 8½. 25237-X Pa. $5.95

SAILING ALONE AROUND THE WORLD, Captain Joshua Slocum. First man to sail around the world, alone, in small boat. One of great feats of seamanship told in delightful manner. 67 illustrations. 294pp. 5⅜ × 8½. 20326-3 Pa. $4.50

LETTERS AND NOTES ON THE MANNERS, CUSTOMS AND CONDITIONS OF THE NORTH AMERICAN INDIANS, George Catlin. Classic account of life among Plains Indians: ceremonies, hunt, warfare, etc. 312 plates. 572pp. of text. 6⅛ × 9¼. 22118-0, 22119-9 Pa. Two-vol. set $15.90

ALASKA: The Harriman Expedition, 1899, John Burroughs, John Muir, et al. Informative, engrossing accounts of two-month, 9,000-mile expedition. Native peoples, wildlife, forests, geography, salmon industry, glaciers, more. Profusely illustrated. 240 black-and-white line drawings. 124 black-and-white photographs. 3 maps. Index. 576pp. 5⅜ × 8½. 25109-8 Pa. $11.95

THE BOOK OF BEASTS: Being a Translation from a Latin Bestiary of the Twelfth Century, T. H. White. Wonderful catalog real and fanciful beasts: manticore, griffin, phoenix, amphivius, jaculus, many more. White's witty erudite commentary on scientific, historical aspects. Fascinating glimpse of medieval mind. Illustrated. 296pp. 5⅜ × 8¼. (Available in U.S. only)   24609-4 Pa. $5.95

FRANK LLOYD WRIGHT: ARCHITECTURE AND NATURE With 160 Illustrations, Donald Hoffmann. Profusely illustrated study of influence of nature—especially prairie—on Wright's designs for Fallingwater, Robie House, Guggenheim Museum, other masterpieces. 96pp. 9¼ × 10¾.   25098-9 Pa. $7.95

FRANK LLOYD WRIGHT'S FALLINGWATER, Donald Hoffmann. Wright's famous waterfall house: planning and construction of organic idea. History of site, owners, Wright's personal involvement. Photographs of various stages of building. Preface by Edgar Kaufmann, Jr. 100 illustrations. 112pp. 9¼ × 10.
23671-4 Pa. $7.95

YEARS WITH FRANK LLOYD WRIGHT: Apprentice to Genius, Edgar Tafel. Insightful memoir by a former apprentice presents a revealing portrait of Wright the man, the inspired teacher, the greatest American architect. 372 black-and-white illustrations. Preface. Index. vi + 228pp. 8¼ × 11.   24801-1 Pa. $9.95

THE STORY OF KING ARTHUR AND HIS KNIGHTS, Howard Pyle. Enchanting version of King Arthur fable has delighted generations with imaginative narratives of exciting adventures and unforgettable illustrations by the author. 41 illustrations. xviii + 313pp. 6⅛ × 9¼.   21445-1 Pa. $5.95

THE GODS OF THE EGYPTIANS, E. A. Wallis Budge. Thorough coverage of numerous gods of ancient Egypt by foremost Egyptologist. Information on evolution of cults, rites and gods; the cult of Osiris; the Book of the Dead and its rites; the sacred animals and birds; Heaven and Hell; and more. 956pp. 6⅛ × 9¼.
22055-9, 22056-7 Pa., Two-vol. set $20.00

A THEOLOGICO-POLITICAL TREATISE, Benedict Spinoza. Also contains unfinished *Political Treatise*. Great classic on religious liberty, theory of government on common consent. R. Elwes translation. Total of 421pp. 5⅜ × 8½.
20249-6 Pa. $6.95

INCIDENTS OF TRAVEL IN CENTRAL AMERICA, CHIAPAS, AND YUCATAN, John L. Stephens. Almost single-handed discovery of Maya culture; exploration of ruined cities, monuments, temples; customs of Indians. 115 drawings. 892pp. 5⅜ × 8½.   22404-X, 22405-8 Pa., Two-vol. set $15.90

LOS CAPRICHOS, Francisco Goya. 80 plates of wild, grotesque monsters and caricatures. Prado manuscript included. 183pp. 6⅛ × 9⅜.   22384-1 Pa. $4.95

AUTOBIOGRAPHY: The Story of My Experiments with Truth, Mohandas K. Gandhi. Not hagiography, but Gandhi in his own words. Boyhood, legal studies, purification, the growth of the Satyagraha (nonviolent protest) movement. Critical, inspiring work of the man who freed India. 480pp. 5⅜ × 8½. (Available in U.S. only)
24593-4 Pa. $6.95

ILLUSTRATED DICTIONARY OF HISTORIC ARCHITECTURE, edited by Cyril M. Harris. Extraordinary compendium of clear, concise definitions for over 5,000 important architectural terms complemented by over 2,000 line drawings. Covers full spectrum of architecture from ancient ruins to 20th-century Modernism. Preface. 592pp. 7½ × 9⅝. 24444-X Pa. $14.95

THE NIGHT BEFORE CHRISTMAS, Clement Moore. Full text, and woodcuts from original 1848 book. Also critical, historical material. 19 illustrations. 40pp. 4⅝ × 6. 22797-9 Pa. $2.25

THE LESSON OF JAPANESE ARCHITECTURE: 165 Photographs, Jiro Harada. Memorable gallery of 165 photographs taken in the 1930's of exquisite Japanese homes of the well-to-do and historic buildings. 13 line diagrams. 192pp. 8⅜ × 11¼. 24778-3 Pa. $8.95

THE AUTOBIOGRAPHY OF CHARLES DARWIN AND SELECTED LETTERS, edited by Francis Darwin. The fascinating life of eccentric genius composed of an intimate memoir by Darwin (intended for his children); commentary by his son, Francis; hundreds of fragments from notebooks, journals, papers; and letters to and from Lyell, Hooker, Huxley, Wallace and Henslow. xi + 365pp. 5⅜ × 8. 20479-0 Pa. $5.95

WONDERS OF THE SKY: Observing Rainbows, Comets, Eclipses, the Stars and Other Phenomena, Fred Schaaf. Charming, easy-to-read poetic guide to all manner of celestial events visible to the naked eye. Mock suns, glories, Belt of Venus, more. Illustrated. 299pp. 5¼ × 8¼. 24402-4 Pa. $7.95

BURNHAM'S CELESTIAL HANDBOOK, Robert Burnham, Jr. Thorough guide to the stars beyond our solar system. Exhaustive treatment. Alphabetical by constellation: Andromeda to Cetus in Vol. 1; Chamaeleon to Orion in Vol. 2; and Pavo to Vulpecula in Vol. 3. Hundreds of illustrations. Index in Vol. 3. 2,000pp. 6⅛ × 9¼. 23567-X, 23568-8, 23673-0 Pa., Three-vol. set $36.85

STAR NAMES: Their Lore and Meaning, Richard Hinckley Allen. Fascinating history of names various cultures have given to constellations and literary and folkloristic uses that have been made of stars. Indexes to subjects. Arabic and Greek names. Biblical references. Bibliography. 563pp. 5⅜ × 8½. 21079-0 Pa. $7.95

THIRTY YEARS THAT SHOOK PHYSICS: The Story of Quantum Theory, George Gamow. Lucid, accessible introduction to influential theory of energy and matter. Careful explanations of Dirac's anti-particles, Bohr's model of the atom, much more. 12 plates. Numerous drawings. 240pp. 5⅜ × 8½. 24895-X Pa. $4.95

CHINESE DOMESTIC FURNITURE IN PHOTOGRAPHS AND MEASURED DRAWINGS, Gustav Ecke. A rare volume, now affordably priced for antique collectors, furniture buffs and art historians. Detailed review of styles ranging from early Shang to late Ming. Unabridged republication. 161 black-and-white drawings, photos. Total of 224pp. 8⅜ × 11¼. (Available in U.S. only) 25171-3 Pa. $12.95

VINCENT VAN GOGH: A Biography, Julius Meier-Graefe. Dynamic, penetrating study of artist's life, relationship with brother, Theo, painting techniques, travels, more. Readable, engrossing. 160pp. 5⅜ × 8½. (Available in U.S. only) 25253-1 Pa. $3.95

HOW TO WRITE, Gertrude Stein. Gertrude Stein claimed anyone could understand her unconventional writing—here are clues to help. Fascinating improvisations, language experiments, explanations illuminate Stein's craft and the art of writing. Total of 414pp. 4⅜ × 6⅜. 23144-5 Pa. $5.95

ADVENTURES AT SEA IN THE GREAT AGE OF SAIL: Five Firsthand Narratives, edited by Elliot Snow. Rare true accounts of exploration, whaling, shipwreck, fierce natives, trade, shipboard life, more. 33 illustrations. Introduction. 353pp. 5⅜ × 8½. 25177-2 Pa. $7.95

THE HERBAL OR GENERAL HISTORY OF PLANTS, John Gerard. Classic descriptions of about 2,850 plants—with over 2,700 illustrations—includes Latin and English names, physical descriptions, varieties, time and place of growth, more. 2,706 illustrations. xlv + 1,678pp. 8½ × 12¼. 23147-X Cloth. $75.00

DOROTHY AND THE WIZARD IN OZ, L. Frank Baum. Dorothy and the Wizard visit the center of the Earth, where people are vegetables, glass houses grow and Oz characters reappear. Classic sequel to *Wizard of Oz*. 256pp. 5⅜ × 8. 24714-7 Pa. $4.95

SONGS OF EXPERIENCE: Facsimile Reproduction with 26 Plates in Full Color, William Blake. This facsimile of Blake's original "Illuminated Book" reproduces 26 full-color plates from a rare 1826 edition. Includes "The Tyger," "London," "Holy Thursday," and other immortal poems. 26 color plates. Printed text of poems. 48pp. 5¼ × 7. 24636-1 Pa. $3.50

SONGS OF INNOCENCE, William Blake. The first and most popular of Blake's famous "Illuminated Books," in a facsimile edition reproducing all 31 brightly colored plates. Additional printed text of each poem. 64pp. 5¼ × 7. 22764-2 Pa. $3.50

PRECIOUS STONES, Max Bauer. Classic, thorough study of diamonds, rubies, emeralds, garnets, etc.: physical character, occurrence, properties, use, similar topics. 20 plates, 8 in color. 94 figures. 659pp. 6⅛ × 9¼. 21910-0, 21911-9 Pa., Two-vol. set $14.90

ENCYCLOPEDIA OF VICTORIAN NEEDLEWORK, S. F. A. Caulfeild and Blanche Saward. Full, precise descriptions of stitches, techniques for dozens of needlecrafts—most exhaustive reference of its kind. Over 800 figures. Total of 679pp. 8⅛ × 11. Two volumes. Vol. 1 22800-2 Pa. $10.95
Vol. 2 22801-0 Pa. $10.95

THE MARVELOUS LAND OF OZ, L. Frank Baum. Second Oz book, the Scarecrow and Tin Woodman are back with hero named Tip, Oz magic. 136 illustrations. 287pp. 5⅜ × 8½. 20692-0 Pa. $5.95

WILD FOWL DECOYS, Joel Barber. Basic book on the subject, by foremost authority and collector. Reveals history of decoy making and rigging, place in American culture, different kinds of decoys, how to make them, and how to use them. 140 plates. 156pp. 7⅞ × 10¾. 20011-6 Pa. $7.95

HISTORY OF LACE, Mrs. Bury Palliser. Definitive, profusely illustrated chronicle of lace from earliest times to late 19th century. Laces of Italy, Greece, England, France, Belgium, etc. Landmark of needlework scholarship. 266 illustrations. 672pp. 6⅛ × 9¼. 24742-2 Pa. $14.95

ILLUSTRATED GUIDE TO SHAKER FURNITURE, Robert Meader. All furniture and appurtenances, with much on unknown local styles. 235 photos. 146pp. 9 × 12.   22819-3 Pa. $7.95

WHALE SHIPS AND WHALING: A Pictorial Survey, George Francis Dow. Over 200 vintage engravings, drawings, photographs of barks, brigs, cutters, other vessels. Also harpoons, lances, whaling guns, many other artifacts. Comprehensive text by foremost authority. 207 black-and-white illustrations. 288pp. 6 × 9.
24808-9 Pa. $8.95

THE BERTRAMS, Anthony Trollope. Powerful portrayal of blind self-will and thwarted ambition includes one of Trollope's most heartrending love stories. 497pp. 5⅜ × 8½.   25119-5 Pa. $8.95

ADVENTURES WITH A HAND LENS, Richard Headstrom. Clearly written guide to observing and studying flowers and grasses, fish scales, moth and insect wings, egg cases, buds, feathers, seeds, leaf scars, moss, molds, ferns, common crystals, etc.—all with an ordinary, inexpensive magnifying glass. 209 exact line drawings aid in your discoveries. 220pp. 5⅜ × 8½.   23330-8 Pa. $3.95

RODIN ON ART AND ARTISTS, Auguste Rodin. Great sculptor's candid, wide-ranging comments on meaning of art; great artists; relation of sculpture to poetry, painting, music; philosophy of life, more. 76 superb black-and-white illustrations of Rodin's sculpture, drawings and prints. 119pp. 8⅜ × 11¼.   24487-3 Pa. $6.95

FIFTY CLASSIC FRENCH FILMS, 1912–1982: A Pictorial Record, Anthony Slide. Memorable stills from Grand Illusion, Beauty and the Beast, Hiroshima, Mon Amour, many more. Credits, plot synopses, reviews, etc. 160pp. 8¼ × 11.
25256-6 Pa. $11.95

THE PRINCIPLES OF PSYCHOLOGY, William James. Famous long course complete, unabridged. Stream of thought, time perception, memory, experimental methods; great work decades ahead of its time. 94 figures. 1,391pp. 5⅜ × 8½.
20381-6, 20382-4 Pa., Two-vol. set $19.90

BODIES IN A BOOKSHOP, R. T. Campbell. Challenging mystery of blackmail and murder with ingenious plot and superbly drawn characters. In the best tradition of British suspense fiction. 192pp. 5⅜ × 8½.   24720-1 Pa. $3.95

CALLAS: PORTRAIT OF A PRIMA DONNA, George Jellinek. Renowned commentator on the musical scene chronicles incredible career and life of the most controversial, fascinating, influential operatic personality of our time. 64 black-and-white photographs. 416pp. 5⅜ × 8¼.   25047-4 Pa. $7.95

GEOMETRY, RELATIVITY AND THE FOURTH DIMENSION, Rudolph Rucker. Exposition of fourth dimension, concepts of relativity as Flatland characters continue adventures. Popular, easily followed yet accurate, profound. 141 illustrations. 133pp. 5⅜ × 8½.   23400-2 Pa. $3.50

HOUSEHOLD STORIES BY THE BROTHERS GRIMM, with pictures by Walter Crane. 53 classic stories—Rumpelstiltskin, Rapunzel, Hansel and Gretel, the Fisherman and his Wife, Snow White, Tom Thumb, Sleeping Beauty, Cinderella, and so much more—lavishly illustrated with original 19th century drawings. 114 illustrations. x + 269pp. 5⅜ × 8½.   21080-4 Pa. $4.50

SUNDIALS, Albert Waugh. Far and away the best, most thorough coverage of ideas, mathematics concerned, types, construction, adjusting anywhere. Over 100 illustrations. 230pp. 5⅜ × 8½. 22947-5 Pa. $4.00

PICTURE HISTORY OF THE NORMANDIE: With 190 Illustrations, Frank O. Braynard. Full story of legendary French ocean liner: Art Deco interiors, design innovations, furnishings, celebrities, maiden voyage, tragic fire, much more. Extensive text. 144pp. 8⅜ × 11¾. 25257-4 Pa. $9.95

THE FIRST AMERICAN COOKBOOK: A Facsimile of "American Cookery," 1796, Amelia Simmons. Facsimile of the first American-written cookbook published in the United States contains authentic recipes for colonial favorites—pumpkin pudding, winter squash pudding, spruce beer, Indian slapjacks, and more. Introductory Essay and Glossary of colonial cooking terms. 80pp. 5⅜ × 8½. 24710-4 Pa. $3.50

101 PUZZLES IN THOUGHT AND LOGIC, C. R. Wylie, Jr. Solve murders and robberies, find out which fishermen are liars, how a blind man could possibly identify a color—purely by your own reasoning! 107pp. 5⅜ × 8½. 20367-0 Pa. $2.00

THE BOOK OF WORLD-FAMOUS MUSIC—CLASSICAL, POPULAR AND FOLK, James J. Fuld. Revised and enlarged republication of landmark work in musico-bibliography. Full information about nearly 1,000 songs and compositions including first lines of music and lyrics. New supplement. Index. 800pp. 5⅜ × 8¼. 24857-7 Pa. $14.95

ANTHROPOLOGY AND MODERN LIFE, Franz Boas. Great anthropologist's classic treatise on race and culture. Introduction by Ruth Bunzel. Only inexpensive paperback edition. 255pp. 5⅜ × 8½. 25245-0 Pa. $5.95

THE TALE OF PETER RABBIT, Beatrix Potter. The inimitable Peter's terrifying adventure in Mr. McGregor's garden, with all 27 wonderful, full-color Potter illustrations. 55pp. 4¼ × 5½. (Available in U.S. only) 22827-4 Pa. $1.75

THREE PROPHETIC SCIENCE FICTION NOVELS, H. G. Wells. *When the Sleeper Wakes, A Story of the Days to Come* and *The Time Machine* (full version). 335pp. 5⅜ × 8½. (Available in U.S. only) 20605-X Pa. $5.95

APICIUS COOKERY AND DINING IN IMPERIAL ROME, edited and translated by Joseph Dommers Vehling. Oldest known cookbook in existence offers readers a clear picture of what foods Romans ate, how they prepared them, etc. 49 illustrations. 301pp. 6⅛ × 9¼. 23563-7 Pa. $6.00

SHAKESPEARE LEXICON AND QUOTATION DICTIONARY, Alexander Schmidt. Full definitions, locations, shades of meaning of every word in plays and poems. More than 50,000 exact quotations. 1,485pp. 6½ × 9¼. 22726-X, 22727-8 Pa., Two-vol. set $27.90

THE WORLD'S GREAT SPEECHES, edited by Lewis Copeland and Lawrence W. Lamm. Vast collection of 278 speeches from Greeks to 1970. Powerful and effective models; unique look at history. 842pp. 5⅜ × 8½. 20468-5 Pa. $10.95

CATALOG OF DOVER BOOKS

THE BLUE FAIRY BOOK, Andrew Lang. The first, most famous collection, with many familiar tales: Little Red Riding Hood, Aladdin and the Wonderful Lamp, Puss in Boots, Sleeping Beauty, Hansel and Gretel, Rumpelstiltskin; 37 in all. 138 illustrations. 390pp. 5⅜ × 8½. 21437-0 Pa. $5.95

THE STORY OF THE CHAMPIONS OF THE ROUND TABLE, Howard Pyle. Sir Launcelot, Sir Tristram and Sir Percival in spirited adventures of love and triumph retold in Pyle's inimitable style. 50 drawings, 31 full-page. xviii + 329pp. 6½ × 9¼. 21883-X Pa. $6.95

AUDUBON AND HIS JOURNALS, Maria Audubon. Unmatched two-volume portrait of the great artist, naturalist and author contains his journals, an excellent biography by his granddaughter, expert annotations by the noted ornithologist, Dr. Elliott Coues, and 37 superb illustrations. Total of 1,200pp. 5⅜ × 8.
Vol. I 25143-8 Pa. $8.95
Vol. II 25144-6 Pa. $8.95

GREAT DINOSAUR HUNTERS AND THEIR DISCOVERIES, Edwin H. Colbert. Fascinating, lavishly illustrated chronicle of dinosaur research, 1820's to 1960. Achievements of Cope, Marsh, Brown, Buckland, Mantell, Huxley, many others. 384pp. 5¼ × 8¼. 24701-5 Pa. $6.95

THE TASTEMAKERS, Russell Lynes. Informal, illustrated social history of American taste 1850's–1950's. First popularized categories Highbrow, Lowbrow, Middlebrow. 129 illustrations. New (1979) afterword. 384pp. 6 × 9.
23993-4 Pa. $6.95

DOUBLE CROSS PURPOSES, Ronald A. Knox. A treasure hunt in the Scottish Highlands, an old map, unidentified corpse, surprise discoveries keep reader guessing in this cleverly intricate tale of financial skullduggery. 2 black-and-white maps. 320pp. 5⅜ × 8½. (Available in U.S. only) 25032-6 Pa. $5.95

AUTHENTIC VICTORIAN DECORATION AND ORNAMENTATION IN FULL COLOR: 46 Plates from "Studies in Design," Christopher Dresser. Superb full-color lithographs reproduced from rare original portfolio of a major Victorian designer. 48pp. 9¼ × 12¼. 25083-0 Pa. $7.95

PRIMITIVE ART, Franz Boas. Remains the best text ever prepared on subject, thoroughly discussing Indian, African, Asian, Australian, and, especially, Northern American primitive art. Over 950 illustrations show ceramics, masks, totem poles, weapons, textiles, paintings, much more. 376pp. 5⅜ × 8. 20025-6 Pa. $6.95

SIDELIGHTS ON RELATIVITY, Albert Einstein. Unabridged republication of two lectures delivered by the great physicist in 1920–21. *Ether and Relativity* and *Geometry and Experience*. Elegant ideas in non-mathematical form, accessible to intelligent layman. vi + 56pp. 5⅜ × 8½. 24511-X Pa. $2.95

THE WIT AND HUMOR OF OSCAR WILDE, edited by Alvin Redman. More than 1,000 ripostes, paradoxes, wisecracks: Work is the curse of the drinking classes, I can resist everything except temptation, etc. 258pp. 5⅜ × 8½. 20602-5 Pa. $3.95

ADVENTURES WITH A MICROSCOPE, Richard Headstrom. 59 adventures with clothing fibers, protozoa, ferns and lichens, roots and leaves, much more. 142 illustrations. 232pp. 5⅜ × 8½. 23471-1 Pa. $3.95

PLANTS OF THE BIBLE, Harold N. Moldenke and Alma L. Moldenke. Standard reference to all 230 plants mentioned in Scriptures. Latin name, biblical reference, uses, modern identity, much more. Unsurpassed encyclopedic resource for scholars, botanists, nature lovers, students of Bible. Bibliography. Indexes. 123 black-and-white illustrations. 384pp. 6 × 9. 25069-5 Pa. $8.95

FAMOUS AMERICAN WOMEN: A Biographical Dictionary from Colonial Times to the Present, Robert McHenry, ed. From Pocahontas to Rosa Parks, 1,035 distinguished American women documented in separate biographical entries. Accurate, up-to-date data, numerous categories, spans 400 years. Indices. 493pp. 6½ × 9¼. 24523-3 Pa. $9.95

THE FABULOUS INTERIORS OF THE GREAT OCEAN LINERS IN HISTORIC PHOTOGRAPHS, William H. Miller, Jr. Some 200 superb photographs capture exquisite interiors of world's great "floating palaces"—1890's to 1980's: *Titanic, Ile de France, Queen Elizabeth, United States, Europa,* more. Approx. 200 black-and-white photographs. Captions. Text. Introduction. 160pp. 8⅞ × 11¾. 24756-2 Pa. $9.95

THE GREAT LUXURY LINERS, 1927–1954: A Photographic Record, William H. Miller, Jr. Nostalgic tribute to heyday of ocean liners. 186 photos of Ile de France, Normandie, Leviathan, Queen Elizabeth, United States, many others. Interior and exterior views. Introduction. Captions. 160pp. 9 × 12. 24056-8 Pa. $9.95

A NATURAL HISTORY OF THE DUCKS, John Charles Phillips. Great landmark of ornithology offers complete detailed coverage of nearly 200 species and subspecies of ducks: gadwall, sheldrake, merganser, pintail, many more. 74 full-color plates, 102 black-and-white. Bibliography. Total of 1,920pp. 8⅝ × 11¼. 25141-1, 25142-X Cloth. Two-vol. set $100.00

THE SEAWEED HANDBOOK: An Illustrated Guide to Seaweeds from North Carolina to Canada, Thomas F. Lee. Concise reference covers 78 species. Scientific and common names, habitat, distribution, more. Finding keys for easy identification. 224pp. 5⅜ × 8½. 25215-9 Pa. $5.95

THE TEN BOOKS OF ARCHITECTURE: The 1755 Leoni Edition, Leon Battista Alberti. Rare classic helped introduce the glories of ancient architecture to the Renaissance. 68 black-and-white plates. 336pp. 8⅜ × 11¼. 25239-6 Pa. $14.95

MISS MACKENZIE, Anthony Trollope. Minor masterpieces by Victorian master unmasks many truths about life in 19th-century England. First inexpensive edition in years. 392pp. 5⅜ × 8½. 25201-9 Pa. $7.95

THE RIME OF THE ANCIENT MARINER, Gustave Doré, Samuel Taylor Coleridge. Dramatic engravings considered by many to be his greatest work. The terrifying space of the open sea, the storms and whirlpools of an unknown ocean, the ice of Antarctica, more—all rendered in a powerful, chilling manner. Full text. 38 plates. 77pp. 9¼ × 12. 22305-1 Pa. $4.95

THE EXPEDITIONS OF ZEBULON MONTGOMERY PIKE, Zebulon Montgomery Pike. Fascinating first-hand accounts (1805–6) of exploration of Mississippi River, Indian wars, capture by Spanish dragoons, much more. 1,088pp. 5⅜ × 8½. 25254-X, 25255-8 Pa. Two-vol. set $23.90

A CONCISE HISTORY OF PHOTOGRAPHY: Third Revised Edition, Helmut Gernsheim. Best one-volume history—camera obscura, photochemistry, daguerreotypes, evolution of cameras, film, more. Also artistic aspects—landscape, portraits, fine art, etc. 281 black-and-white photographs. 26 in color. 176pp. 8⅜ × 11¼. 25128-4 Pa. $12.95

THE DORÉ BIBLE ILLUSTRATIONS, Gustave Doré. 241 detailed plates from the Bible: the Creation scenes, Adam and Eve, Flood, Babylon, battle sequences, life of Jesus, etc. Each plate is accompanied by the verses from the King James version of the Bible. 241pp. 9 × 12. 23004-X Pa. $8.95

HUGGER-MUGGER IN THE LOUVRE, Elliot Paul. Second Homer Evans mystery-comedy. Theft at the Louvre involves sleuth in hilarious, madcap caper. "A knockout."—Books. 336pp. 5⅜ × 8½. 25185-3 Pa. $5.95

FLATLAND, E. A. Abbott. Intriguing and enormously popular science-fiction classic explores the complexities of trying to survive as a two-dimensional being in a three-dimensional world. Amusingly illustrated by the author. 16 illustrations. 103pp. 5⅜ × 8½. 20001-9 Pa. $2.00

THE HISTORY OF THE LEWIS AND CLARK EXPEDITION, Meriwether Lewis and William Clark, edited by Elliott Coues. Classic edition of Lewis and Clark's day-by-day journals that later became the basis for U.S. claims to Oregon and the West. Accurate and invaluable geographical, botanical, biological, meteorological and anthropological material. Total of 1,508pp. 5⅜ × 8½.
21268-8, 21269-6, 21270-X Pa. Three-vol. set $25.50

LANGUAGE, TRUTH AND LOGIC, Alfred J. Ayer. Famous, clear introduction to Vienna, Cambridge schools of Logical Positivism. Role of philosophy, elimination of metaphysics, nature of analysis, etc. 160pp. 5⅜ × 8½. (Available in U.S. and Canada only) 20010-8 Pa. $2.95

MATHEMATICS FOR THE NONMATHEMATICIAN, Morris Kline. Detailed, college-level treatment of mathematics in cultural and historical context, with numerous exercises. For liberal arts students. Preface. Recommended Reading Lists. Tables. Index. Numerous black-and-white figures. xvi + 641pp. 5⅜ × 8½.
24823-2 Pa. $11.95

28 SCIENCE FICTION STORIES, H. G. Wells. Novels, *Star Begotten* and *Men Like Gods,* plus 26 short stories: "Empire of the Ants," "A Story of the Stone Age," "The Stolen Bacillus," "In the Abyss," etc. 915pp. 5⅜ × 8½. (Available in U.S. only)
20265-8 Cloth. $10.95

HANDBOOK OF PICTORIAL SYMBOLS, Rudolph Modley. 3,250 signs and symbols, many systems in full; official or heavy commercial use. Arranged by subject. Most in Pictorial Archive series. 143pp. 8⅛ × 11. 23357-X Pa. $5.95

INCIDENTS OF TRAVEL IN YUCATAN, John L. Stephens. Classic (1843) exploration of jungles of Yucatan, looking for evidences of Maya civilization. Travel adventures, Mexican and Indian culture, etc. Total of 669pp. 5⅜ × 8½.
20926-1, 20927-X Pa., Two-vol. set $9.90

DEGAS: An Intimate Portrait, Ambroise Vollard. Charming, anecdotal memoir by famous art dealer of one of the greatest 19th-century French painters. 14 black-and-white illustrations. Introduction by Harold L. Van Doren. 96pp. 5⅜ × 8½.
25131-4 Pa. $3.95

PERSONAL NARRATIVE OF A PILGRIMAGE TO ALMANDINAH AND MECCAH, Richard Burton. Great travel classic by remarkably colorful personality. Burton, disguised as a Moroccan, visited sacred shrines of Islam, narrowly escaping death. 47 illustrations. 959pp. 5⅜ × 8½.   21217-3, 21218-1 Pa., Two-vol. set $17.90

PHRASE AND WORD ORIGINS, A. H. Holt. Entertaining, reliable, modern study of more than 1,200 colorful words, phrases, origins and histories. Much unexpected information. 254pp. 5⅜ × 8½.   20758-7 Pa. $4.95

THE RED THUMB MARK, R. Austin Freeman. In this first Dr. Thorndyke case, the great scientific detective draws fascinating conclusions from the nature of a single fingerprint. Exciting story, authentic science. 320pp. 5⅜ × 8½. (Available in U.S. only)   25210-8 Pa. $5.95

AN EGYPTIAN HIEROGLYPHIC DICTIONARY, E. A. Wallis Budge. Monumental work containing about 25,000 words or terms that occur in texts ranging from 3000 B.C. to 600 A.D. Each entry consists of a transliteration of the word, the word in hieroglyphs, and the meaning in English. 1,314pp. 6⅜ × 10.
23615-3, 23616-1 Pa., Two-vol. set $27.90

THE COMPLEAT STRATEGYST: Being a Primer on the Theory of Games of Strategy, J. D. Williams. Highly entertaining classic describes, with many illustrated examples, how to select best strategies in conflict situations. Prefaces. Appendices. xvi + 268pp. 5⅜ × 8½.   25101-2 Pa. $5.95

THE ROAD TO OZ, L. Frank Baum. Dorothy meets the Shaggy Man, little Button-Bright and the Rainbow's beautiful daughter in this delightful trip to the magical Land of Oz. 272pp. 5⅜ × 8.   25208-6 Pa. $4.95

POINT AND LINE TO PLANE, Wassily Kandinsky. Seminal exposition of role of point, line, other elements in non-objective painting. Essential to understanding 20th-century art. 127 illustrations. 192pp. 6½ × 9¼.   23808-3 Pa. $4.50

LADY ANNA, Anthony Trollope. Moving chronicle of Countess Lovel's bitter struggle to win for herself and daughter Anna their rightful rank and fortune— perhaps at cost of sanity itself. 384pp. 5⅜ × 8½.   24669-8 Pa. $6.95

EGYPTIAN MAGIC, E. A. Wallis Budge. Sums up all that is known about magic in Ancient Egypt: the role of magic in controlling the gods, powerful amulets that warded off evil spirits, scarabs of immortality, use of wax images, formulas and spells, the secret name, much more. 253pp. 5⅜ × 8½.   22681-6 Pa. $4.00

THE DANCE OF SIVA, Ananda Coomaraswamy. Preeminent authority unfolds the vast metaphysic of India: the revelation of her art, conception of the universe, social organization, etc. 27 reproductions of art masterpieces. 192pp. 5⅜ × 8½.
24817-8 Pa. $5.95

CHRISTMAS CUSTOMS AND TRADITIONS, Clement A. Miles. Origin, evolution, significance of religious, secular practices. Caroling, gifts, yule logs, much more. Full, scholarly yet fascinating; non-sectarian. 400pp. 5⅜ × 8½.
23354-5 Pa. $6.50

THE HUMAN FIGURE IN MOTION, Eadweard Muybridge. More than 4,500 stopped-action photos, in action series, showing undraped men, women, children jumping, lying down, throwing, sitting, wrestling, carrying, etc. 390pp. 7⅞ × 10⅝.
20204-6 Cloth. $19.95

THE MAN WHO WAS THURSDAY, Gilbert Keith Chesterton. Witty, fast-paced novel about a club of anarchists in turn-of-the-century London. Brilliant social, religious, philosophical speculations. 128pp. 5⅜ × 8½. 25121-7 Pa. $3.95

A CEZANNE SKETCHBOOK: Figures, Portraits, Landscapes and Still Lifes, Paul Cezanne. Great artist experiments with tonal effects, light, mass, other qualities in over 100 drawings. A revealing view of developing master painter, precursor of Cubism. 102 black-and-white illustrations. 144pp. 8¾ × 6⅜. 24790-2 Pa. $5.95

AN ENCYCLOPEDIA OF BATTLES: Accounts of Over 1,560 Battles from 1479 B.C. to the Present, David Eggenberger. Presents essential details of every major battle in recorded history, from the first battle of Megiddo in 1479 B.C. to Grenada in 1984. List of Battle Maps. New Appendix covering the years 1967–1984. Index. 99 illustrations. 544pp. 6½ × 9¼. 24913-1 Pa. $14.95

AN ETYMOLOGICAL DICTIONARY OF MODERN ENGLISH, Ernest Weekley. Richest, fullest work, by foremost British lexicographer. Detailed word histories. Inexhaustible. Total of 856pp. 6½ × 9¼.
21873-2, 21874-0 Pa., Two-vol. set $17.00

WEBSTER'S AMERICAN MILITARY BIOGRAPHIES, edited by Robert McHenry. Over 1,000 figures who shaped 3 centuries of American military history. Detailed biographies of Nathan Hale, Douglas MacArthur, Mary Hallaren, others. Chronologies of engagements, more. Introduction. Addenda. 1,033 entries in alphabetical order. xi + 548pp. 6½ × 9¼. (Available in U.S. only)
24758-9 Pa. $11.95

LIFE IN ANCIENT EGYPT, Adolf Erman. Detailed older account, with much not in more recent books: domestic life, religion, magic, medicine, commerce, and whatever else needed for complete picture. Many illustrations. 597pp. 5⅜ × 8½.
22632-8 Pa. $8.50

HISTORIC COSTUME IN PICTURES, Braun & Schneider. Over 1,450 costumed figures shown, covering a wide variety of peoples: kings, emperors, nobles, priests, servants, soldiers, scholars, townsfolk, peasants, merchants, courtiers, cavaliers, and more. 256pp. 8⅜ × 11¼. 23150-X Pa. $7.95

THE NOTEBOOKS OF LEONARDO DA VINCI, edited by J. P. Richter. Extracts from manuscripts reveal great genius; on painting, sculpture, anatomy, sciences, geography, etc. Both Italian and English. 186 ms. pages reproduced, plus 500 additional drawings, including studies for *Last Supper, Sforza* monument, etc. 860pp. 7⅞ × 10¾. (Available in U.S. only) 22572-0, 22573-9 Pa., Two-vol. set $25.90

THE ART NOUVEAU STYLE BOOK OF ALPHONSE MUCHA: All 72 Plates from "Documents Decoratifs" in Original Color, Alphonse Mucha. Rare copyright-free design portfolio by high priest of Art Nouveau. Jewelry, wallpaper, stained glass, furniture, figure studies, plant and animal motifs, etc. Only complete one-volume edition. 80pp. 9⅜ × 12¼. 24044-4 Pa. $8.95

ANIMALS: 1,419 COPYRIGHT-FREE ILLUSTRATIONS OF MAMMALS, BIRDS, FISH, INSECTS, ETC., edited by Jim Harter. Clear wood engravings present, in extremely lifelike poses, over 1,000 species of animals. One of the most extensive pictorial sourcebooks of its kind. Captions. Index. 284pp. 9 × 12. 23766-4 Pa. $9.95

OBELISTS FLY HIGH, C. Daly King. Masterpiece of American detective fiction, long out of print, involves murder on a 1935 transcontinental flight—"a very thrilling story"—NY Times. Unabridged and unaltered republication of the edition published by William Collins Sons & Co. Ltd., London, 1935. 288pp. 5⅜ × 8½. (Available in U.S. only) 25036-9 Pa. $4.95

VICTORIAN AND EDWARDIAN FASHION: A Photographic Survey, Alison Gernsheim. First fashion history completely illustrated by contemporary photographs. Full text plus 235 photos, 1840–1914, in which many celebrities appear. 240pp. 6½ × 9¼. 24205-6 Pa. $6.00

THE ART OF THE FRENCH ILLUSTRATED BOOK, 1700–1914, Gordon N. Ray. Over 630 superb book illustrations by Fragonard, Delacroix, Daumier, Doré, Grandville, Manet, Mucha, Steinlen, Toulouse-Lautrec and many others. Preface. Introduction. 633 halftones. Indices of artists, authors & titles, binders and provenances. Appendices. Bibliography. 608pp. 8⅜ × 11¼. 25086-5 Pa. $24.95

THE WONDERFUL WIZARD OF OZ, L. Frank Baum. Facsimile in full color of America's finest children's classic. 143 illustrations by W. W. Denslow. 267pp. 5⅜ × 8½. 20691-2 Pa. $5.95

FRONTIERS OF MODERN PHYSICS: New Perspectives on Cosmology, Relativity, Black Holes and Extraterrestrial Intelligence, Tony Rothman, et al. For the intelligent layman. Subjects include: cosmological models of the universe; black holes; the neutrino; the search for extraterrestrial intelligence. Introduction. 46 black-and-white illustrations. 192pp. 5⅜ × 8½. 24587-X Pa. $6.95

THE FRIENDLY STARS, Martha Evans Martin & Donald Howard Menzel. Classic text marshalls the stars together in an engaging, non-technical survey, presenting them as sources of beauty in night sky. 23 illustrations. Foreword. 2 star charts. Index. 147pp. 5⅜ × 8½. 21099-5 Pa. $3.50

FADS AND FALLACIES IN THE NAME OF SCIENCE, Martin Gardner. Fair, witty appraisal of cranks, quacks, and quackeries of science and pseudoscience: hollow earth, Velikovsky, orgone energy, Dianetics, flying saucers, Bridey Murphy, food and medical fads, etc. Revised, expanded In the Name of Science. "A very able and even-tempered presentation."—The New Yorker. 363pp. 5⅜ × 8. 20394-8 Pa. $5.95

ANCIENT EGYPT: ITS CULTURE AND HISTORY, J. E Manchip White. From pre-dynastics through Ptolemies: society, history, political structure, religion, daily life, literature, cultural heritage. 48 plates. 217pp. 5⅜ × 8½. 22548-8 Pa. $4.95

SIR HARRY HOTSPUR OF HUMBLETHWAITE, Anthony Trollope. Incisive, unconventional psychological study of a conflict between a wealthy baronet, his idealistic daughter, and their scapegrace cousin. The 1870 novel in its first inexpensive edition in years. 250pp. 5⅜ × 8½. 24953-0 Pa. $4.95

LASERS AND HOLOGRAPHY, Winston E. Kock. Sound introduction to burgeoning field, expanded (1981) for second edition. Wave patterns, coherence, lasers, diffraction, zone plates, properties of holograms, recent advances. 84 illustrations. 160pp. 5⅜ × 8¼. (Except in United Kingdom) 24041-X Pa. $3.50

INTRODUCTION TO ARTIFICIAL INTELLIGENCE: SECOND, EN-LARGED EDITION, Philip C. Jackson, Jr. Comprehensive survey of artificial intelligence—the study of how machines (computers) can be made to act intelli-gently. Includes introductory and advanced material. Extensive notes updating the main text. 132 black-and-white illustrations. 512pp. 5⅜ × 8½. 24864-X Pa. $8.95

HISTORY OF INDIAN AND INDONESIAN ART, Ananda K. Coomaraswamy. Over 400 illustrations illuminate classic study of Indian art from earliest Harappa finds to early 20th century. Provides philosophical, religious and social insights. 304pp. 6⅜ × 9⅜. 25005-9 Pa. $8.95

THE GOLEM, Gustav Meyrink. Most famous supernatural novel in modern European literature, set in Ghetto of Old Prague around 1890. Compelling story of mystical experiences, strange transformations, profound terror. 13 black-and-white illustrations. 224pp. 5⅜ × 8½. (Available in U.S. only) 25025-3 Pa. $5.95

ARMADALE, Wilkie Collins. Third great mystery novel by the author of *The Woman in White* and *The Moonstone*. Original magazine version with 40 illustrations. 597pp. 5⅜ × 8½. 23429-0 Pa. $7.95

PICTORIAL ENCYCLOPEDIA OF HISTORIC ARCHITECTURAL PLANS, DETAILS AND ELEMENTS: With 1,880 Line Drawings of Arches, Domes, Doorways, Facades, Gables, Windows, etc., John Theodore Haneman. Sourcebook of inspiration for architects, designers, others. Bibliography. Captions. 141pp. 9 × 12. 24605-1 Pa. $6.95

BENCHLEY LOST AND FOUND, Robert Benchley. Finest humor from early 30's, about pet peeves, child psychologists, post office and others. Mostly unavailable elsewhere. 73 illustrations by Peter Arno and others. 183pp. 5⅜ × 8½. 22410-4 Pa. $3.95

ERTÉ GRAPHICS, Erté. Collection of striking color graphics: *Seasons, Alphabet, Numerals, Aces* and *Precious Stones*. 50 plates, including 4 on covers. 48pp. 9⅜ × 12¼. 23580-7 Pa. $6.95

THE JOURNAL OF HENRY D. THOREAU, edited by Bradford Torrey, F. H. Allen. Complete reprinting of 14 volumes, 1837–61, over two million words; the sourcebooks for *Walden*, etc. Definitive. All original sketches, plus 75 photographs. 1,804pp. 8½ × 12¼. 20312-3, 20313-1 Cloth., Two-vol. set $80.00

CASTLES: THEIR CONSTRUCTION AND HISTORY, Sidney Toy. Traces castle development from ancient roots. Nearly 200 photographs and drawings illustrate moats, keeps, baileys, many other features. Caernarvon, Dover Castles, Hadrian's Wall, Tower of London, dozens more. 256pp. 5⅜ × 8¼.
24898-4 Pa. $5.95

AMERICAN CLIPPER SHIPS: 1833–1858, Octavius T. Howe & Frederick C. Matthews. Fully-illustrated, encyclopedic review of 352 clipper ships from the period of America's greatest maritime supremacy. Introduction. 109 halftones. 5 black-and-white line illustrations. Index. Total of 928pp. 5⅜ × 8½.
25115-2, 25116-0 Pa., Two-vol. set $17.90

TOWARDS A NEW ARCHITECTURE, Le Corbusier. Pioneering manifesto by great architect, near legendary founder of "International School." Technical and aesthetic theories, views on industry, economics, relation of form to function, "mass-production spirit," much more. Profusely illustrated. Unabridged translation of 13th French edition. Introduction by Frederick Etchells. 320pp. 6⅛ × 9¼.
(Available in U.S. only) 25023-7 Pa. $8.95

THE BOOK OF KELLS, edited by Blanche Cirker. Inexpensive collection of 32 full-color, full-page plates from the greatest illuminated manuscript of the Middle Ages, painstakingly reproduced from rare facsimile edition. Publisher's Note. Captions. 32pp. 9⅜ × 12¼. 24345-1 Pa. $4.50

BEST SCIENCE FICTION STORIES OF H. G. WELLS, H. G. Wells. Full novel *The Invisible Man,* plus 17 short stories: "The Crystal Egg," "Aepyornis Island," "The Strange Orchid," etc. 303pp. 5⅜ × 8½. (Available in U.S. only)
21531-8 Pa. $4.95

AMERICAN SAILING SHIPS: Their Plans and History, Charles G. Davis. Photos, construction details of schooners, frigates, clippers, other sailcraft of 18th to early 20th centuries—plus entertaining discourse on design, rigging, nautical lore, much more. 137 black-and-white illustrations. 240pp. 6⅛ × 9¼.
24658-2 Pa. $5.95

ENTERTAINING MATHEMATICAL PUZZLES, Martin Gardner. Selection of author's favorite conundrums involving arithmetic, money, speed, etc., with lively commentary. Complete solutions. 112pp. 5⅜ × 8½. 25211-6 Pa. $2.95
THE WILL TO BELIEVE, HUMAN IMMORTALITY, William James. Two books bound together. Effect of irrational on logical, and arguments for human immortality. 402pp. 5⅜ × 8½. 20291-7 Pa. $7.50

THE HAUNTED MONASTERY and THE CHINESE MAZE MURDERS, Robert Van Gulik. 2 full novels by Van Gulik continue adventures of Judge Dee and his companions. An evil Taoist monastery, seemingly supernatural events; overgrown topiary maze that hides strange crimes. Set in 7th-century China. 27 illustrations. 328pp. 5⅜ × 8½. 23502-5 Pa. $5.00

CELEBRATED CASES OF JUDGE DEE (DEE GOONG AN), translated by Robert Van Gulik. Authentic 18th-century Chinese detective novel; Dee and associates solve three interlocked cases. Led to Van Gulik's own stories with same characters. Extensive introduction. 9 illustrations. 237pp. 5⅜ × 8½.
23337-5 Pa. $4.95

Prices subject to change without notice.
Available at your book dealer or write for free catalog to Dept. GI, Dover Publications, Inc., 31 East 2nd St., Mineola, N.Y. 11501. Dover publishes more than 175 books each year on science, elementary and advanced mathematics, biology, music, art, literary history, social sciences and other areas.